SpringerBriefs in Electrical and Computer Engineering

W0079536

For further volumes:
http://www.springer.com/series/10059

SpringerBriefs in Electrical and Computer Engineering

Cam Nguyen • Jeongwoo Han

Time-Domain Ultra-Wideband Radar, Sensor and Components

Theory, Analysis and Design

 Springer

Cam Nguyen
Texas A&M University Dept of Elec
    and Comp Engineering
College Station
Texas
USA

Jeongwoo Han
Radio Engineering Center
Pine-Telecom Inc.
Yuseong-gu, Daejeon
Korea, Republic of (South Korea)

ISSN 2191-8112   ISSN 2191-8120 (electronic)
ISBN 978-1-4614-9577-2   ISBN 978-1-4614-9578-9 (eBook)
DOI 10.1007/978-1-4614-9578-9
Springer New York Heidelberg Dordrecht London

Library of Congress Control Number: 2014932677

Printed on acid-free paper

Springer is part of Springer Science+Business Media (www.springer.com)

Preface

Impulse-based ultra-wideband (UWB) system or, in short, UWB system is a unique system possessing many desired characteristics, owing to the transmission and reception of only a single signal having pulse waveform at all times, rather than multiple consecutive CW signals having sinusoidal waveform at different times as in CW-based systems. This unique operation is equivalent to transmitting and receiving many CW signals across an extremely broad bandwidth concurrently. In other words, an UWB system is electrically equivalent to multiple CW systems, each operating at a single frequency in an ultra-wide bandwidth, working simultaneously. It is this unique operation that makes the UWB system distinctly different from the CW systems, not only in the design and performance, but also in some applications—some of which are not even possible with CW systems. UWB systems find numerous applications for military, security, civilian, commerce, and medicine. UWB systems are suitable for both high resolution and long range, but they are particularly attractive for high-resolution sensing applications.

This book is devoted to the theory, analysis and design of UWB systems and their components. Particularly, it covers the main topics of UWB systems that are system analysis, transmitter design, receiver design, antenna design, and system integration and test. It also presents the design of a specific practical UWB system and its constituent components of transmitter, receiver, antenna, signal processing, integration, and test, which serve as an effective way to demonstrate not only the analysis, design, and applications of UWB systems, but also the analysis and design of constituent components. Although the book is succinct, the material is very much self-contained and contains practical, valuable and sufficient information presented in such a way that allows readers with an undergraduate background in electrical engineering or physics, with some experiences or graduate courses in microwave circuits, to understand and design easily UWB components, transmitters, receivers, and systems for various applications.

The book is useful for engineers, physicists, and graduate students who work in radar, sensor, and communication systems as well as those involved in the design of microwave circuits and systems. It is our sincere hope that the book can serve not only as a reference for the development of UWB systems and components, but also for a new generation of innovative ideas that can benefit many existing sensing and communication applications or be implemented for other new applications.

College Station, Texas; Newport Beach, California, U.S.A.   Cam Nguyen
Seoul, South Korea   Jeongwoo Han

Contents

Chapter 1
Introduction

The origin of "Ultra-Wideband" technology can be traced back to the early 1960's on time-domain electromagnetics, when the study of electromagnetic-wave propagation was primarily viewed from the time-domain perspective, rather than from the more common frequency domain. Nevertheless, the term Ultra-Wideband was used for the first time in 1994 by the U.S. Department of Defense to indicate impulse-based systems. In general, however, ultra-wideband (UWB) systems should refer to systems that operate over an ultra-wide bandwidth. These systems may have different architectures and are used for different applications such as impulse-based systems and continuous-wave (CW) based systems with different transmitting signal's frequency modulations like frequency modulated continuous wave (FMCW) or stepped-frequency system. Specifically, a system may be classified as an UWB system when its operating frequency range is wider than 500 MHz or 20 % fractional bandwidth. The most distinguished difference between an impulse system and a FMCW or stepped-frequency system (and in general other non-impulse or CW based systems) is in the transmitting waveform. An FMCW system transmits and receives CW (sinusoidal) signals, one signal at each frequency, subsequently across a bandwidth. An FMCW system does not transmit and receive signals of different frequencies simultaneously. That is, an FMCW system is basically operated over a bandwidth of single-frequency signals. A stepped-frequency system transmits consecutive trains of CW signals, each at different frequency separated by a certain amount, toward targets and receives reflected signals from the targets. The received digital in-phase and quadrature signals are then transformed into a synthetic pulse in time domain using inverse discrete Fourier Transform. A stepped-frequency system has very narrow instantaneous bandwidth at each frequency, resulting in high signal-to-noise ratio at the receiver. Its entire bandwidth, on the other hand, can be very wide, leading to fine resolution. Moreover, its high average transmitting power enables deep penetration or long range. Although the final received signal is transformed into a time-domain pulse signal, it is still CW-based and does not transmit and receive signals of different frequencies simultaneously. On the other hand, an impulse system transmits and receives a periodic (non-sinusoidal) impulse-type signal which contains many constituent signals occurring simultaneously, each having a different frequency. In other words, an impulse system transmits and receives many CW signals having different

C. Nguyen, J. Han, *Time-Domain Ultra-Wideband Radar, Sensor and Components*,
SpringerBriefs in Electrical and Computer Engineering,
DOI 10.1007/978-1-4614-9578-9_1, © Springer International Publishing Switzerland 2014

frequencies concurrently. It is this characteristic that makes the impulse system and FMCW or stepped-frequency system (and other CW based systems) distinctively different in the design, operation, performance, and possible applications.

It is especially noted that this book only covers the (time-domain) impulse-based UWB systems, and so the name UWB systems hereafter specifically imply (time-domain) impulse systems. Compared to CW based systems, UWB systems have many advantages as following:

1.1 Fine Resolution and Long Range

UWB systems typically have much wider instantaneous bandwidths than those of CW based systems due to the extremely wideband nature of the impulse-type signals. These signals contain both low and high frequency components, making impulse systems very suitable for applications requiring fine range-resolution and/ or long range. An ultra-wide bandwidth directly leads to fine range-resolution due to the fact that the range resolution is inversely proportional to the bandwidth. An ultra-wide frequency range spanning across low and high frequencies enables long range due to small attenuation at low frequencies and hence long propagating distance of low-frequency signals. It is noted that, while similar range-resolution and range can also be achieved for CW based systems operating with same bandwidth and frequencies, respectively, it is very difficult (if not impossible) to realize an extremely wide-band CW system.

1.2 High Multi-Path Resolution and Low Interference with Other Existing Signals

The transmitting energy spectral density of UWB systems is much lower than that of CW based systems for the same input power since the total energy is spread over a wide range of frequencies of the impulse signal. This effectively produces much smaller interference to the signals of other existing or co-operating RF systems. Typical impulse signals have a very narrow pulse width and hence the transmission duration of impulse signals is very short in most cases. The signals returned from targets, in turn, have a very short time-window of opportunity to collide with the line-of-sight signals and less likely cause signal degradation, hence very high multi-path resolution can be achieved. The large frequency diversity of the ultra-wide spectrum of an impulse signal makes impulse signal relatively resistant to intentional and unintentional jamming or interference, because it is difficult to jam every frequency component in the ultra-wideband spectrum at once. Even if some of the frequency components are jammed, there is still possibly a large range of frequencies that remains untouched. UWB systems offer excellent immunity to interference from other existing signals, while also causing minimum interference to these signals.

1.3   Low Probability of Interception or Detection

The lower-energy spectrum density of impulse signals also makes unintended detection more difficult than CW systems, hence resulting in low probability of interception or detection, which is desirable for secure and military applications.

1.4   Reduced Signal Diminishing

The ultra-wide bandwidth of UWB systems leads to high frequency diversity that reduces the chance of signal diminishing in certain operating environments with severe multi-path fading at some frequencies, such as indoors, urban settings or mountainous terrains, or when signal attenuation at some frequencies are excessively high, which hinder the sensing capabilities, or where noise exists in a narrow-frequency range within the operating band, resulting in better immunity to destructive environments.

1.5   Reduced Signal Diminishing

The accuracy of locating and tracking for UWB systems is higher due to the narrow time-duration of pulse signals typically used for transmission which causes much better timing precision than CW systems such as global positioning system (GPS).

1.6   Simple and Low-Cost Architecture

UWB systems can be implemented with a simpler architecture than their wideband CW counterparts, in which the core signal generator of the transmitter can be realized with a simple pulse generator without an up-conversion circuit and the core mixer of the receiver can be implemented with a simple direct-conversion (to baseband) sampling circuit, that does not require an intermediate frequency-conversion stage, as compared to more complicated wideband signal source and mixer typically used in CW based systems. Moreover, a complex frequency synthesizer needed for the transmitter and receiver in CW systems is avoided.

Needless to say, UWB systems indeed have several disadvantages as compared to CW systems. For instance, the receiver's noise figure of UWB systems is much higher than that of CW systems, which in turn limits the sensitivity and hence the dynamic range of the receiver, preventing their use for applications requiring very high sensitivity and/or large dynamic range. It is also much more complicated to design antennas for UWB systems due to wide bandwidths and the need to maintain signal fidelity across such bandwidths.

UWB radar and sensor systems find numerous applications for military, security, civilian, commerce and medicine, and various UWB radar and sensor systems have been developed, e.g., [1–12]. While UWB systems are suitable for both high resolution and long range, as mentioned earlier, they are particularly attractive and dominant for high-resolution sensing applications. The following are some of the existing and emerging applications of UWB radar and sensor systems:

Military and Security Applications: Detection, location and identification of targets such as aircrafts, tunnels, concealed weapons, hidden illegal drugs, buried mine and unexploded ordnance (UXO); locating and tracking personnel; detection and identification of hidden activities; access control; through-wall imaging and surveillance; building surveillance and monitoring.

Civilian and Commercial Applications: Detection, identification and assessment of abnormal conditions of civil structures such as pavements, bridges, buildings, buried underground pipes; detection, location and identification of objects; asset and inventory management; radio-frequency identification (RFID); monitoring of personal properties such as cars, homes and valuable items; intrusion detection; asset tracking; measurement of liquid volumes and levels; inspection, evaluation and process control of materials; geophysical prospecting, altimetry; collision and obstacle avoidance for automobile and aviation

Medical Applications: Detection and imaging of tumors; health monitoring of elders; health examination of patients; medical imaging.

There are many aspects in UWB radar and sensor systems and a complete coverage would require a book of substantial volume. The objective of this book is not to provide a full coverage of UWB radar and sensor systems. It is our intention to address only the essential parts of UWB radar and sensor including system and component design, analysis, and measurement in a concise manner, yet with sufficient details to allow the readers to understand and design UWB systems and components for their intended applications, whether for research or for commercial usage.

This book presents the theory, analysis and design of UWB radar and sensor systems and their components. It addresses five main topics of UWB systems: system analysis, transmitter design, receiver design, antenna design, and system integration and test. The developments of a practical UWB system and its components using microwave integrated circuits (MICs), as well as various measurements, are included in details to demonstrate the theory, analysis, design, and some applications.

The book is organized as follows. Chapter 1 gives the introduction of UWB systems and their possible applications. Chapter 2 addresses the system analysis including power budget and range resolution. Chapter 3 describes the transmitter design including the design, fabrication and measurement of impulse and monocycle pulse generators. Chapter 4 covers the receiver design where the design, fabrication and measurement of strobe pulse generator, sampler, low-noise amplifier, and synchronous sampling receiver are addressed. Chapter 5 presents the design, fabrication and measurement of two ultra-wideband antennas including the microstrip quasi-horn antenna and the UWB uniplanar antenna. Chapter 6 is devoted to the system integration, signal processing and measurement. Finally, Chap. 7 gives the summary and conclusion.

References

1. Skolnik, M.I.: An introduction to impulse radar. Naval Research Laboratory, Washington, DC, NRL Memorandum Report 6755 (Nov 1990)
2. Daniels, D.J.: Surface Penetrating Radar. IEE, London (1996)
3. Daniels, D.J., Gunton, D.J., Scott, H.F.: Introduction to subsurface radar. IEE Proc. **135**(4), 278–320 (1988)
4. Taylor, J.D.: Introduction to Ultra-Wideband Radar Systems. CRC, Boca Raton (1995)
5. Taylor, J.D.: Ultra-Wideband Radar Technology. CRC, Boca Raton (2001)
6. Park, J.S., Nguyen, C.: An ultra-wideband microwave radar sensor for nondestructive evaluation of pavement subsurface. IEEE Sens. J. **5**, 942–949 (2005)
7. Fontana, R.J.: Recent applications of ultra wideband radar and communications systems. In: Smith, P.D., Cloude, S.R. (eds) Ultra-Wideband, Short-Pulse Electromagnetics 5, pp. 225–234. Kluwer Academic/Plenum, New York (2002)
8. Yarovoy, A., Ligthart, L.: Full-polarimetric video impulse radar for landmine detection: Experimental verification of main design ideas. In: Yarovoy, A. (eds.) Proc. 2nd Int. Workshop on Advanced Ground Penetrating Radar, pp. 148–155. The International Research Centre for Telecommunicationstransmissions and Radar (IRCTR), Delft (2003)
9. Lee, J.S., Nguyen, C., Scullion, T.: A novel compact, low-cost impulse ground penetrating radar for nondestructive evaluation of pavements. IEEE Trans. Instrum. Meas. **53**, 1502–1509 (2004)
10. Warhus, J.P.: Advanced ground-penetrating, imaging radar for bridge inspection. Lawrence Livermore National Laboratory, CA. (Sept 1994). http://www-eng.llnl.gov/dsed/documents/em/jwpctta93.html
11. Azevedo, S., McEwan, T.E.: Micropower impulse radar. Sci. Tech. Rev. 17–29 (Jan/Feb 1996)
12. Han, J.W., Nguyen, C.: Development of a tunable multi-band UWB radar sensor and its applications to subsurface sensing. IEEE Sens. J. **7**(1), 51–58 (Jan. 2007)

Chapter 2
System Analysis

2.1 Introduction

Two of the fundamental questions in the design of impulse-based UWB systems, or simply UWB systems as referred to in this book, are how much transmitting power is needed and what resolution is required to distinguish different targets. For UWB applications involving targets of multiple stratified media, another important question likely to be raised is how the radiated electromagnetic (EM) wave or signal propagates in the stratified media and what effect reflections have on received signals. In this chapter, a system analysis is conducted to find answers for these problems for the design of UWB systems. For illustration purposes, the analysis assumes a multi-layer structure as a specific target to be sensed by a UWB system. The analysis results will be used as reference specifications for the design of a UWB system to be described in subsequent chapters including transmitter, receiver, and antenna.

Specifically, the purposes of the system analysis performed here are to make rough estimations of the required power budget (i.e., required transmitting power) and range resolution for UWB systems. This simple analysis avoids the need of delving deeply into more complicated analysis and design of UWB systems which require more accurate information for the targets and more sophisticated modeling of the transmission and reflection of EM waves in multi-layer structures. More general and in-depth analysis of UWB systems can be found in the literature. The power budget analysis is based on the method of factorization of total loss presented in [1]. The resolution is determined predominantly by the radiating pulse duration, and hence an estimation of the required minimum pulse duration is derived based on the minimum thickness among the layers of a multi-layer structure.

2.2 UWB System Operation

As mentioned in Chap. 1, there are various applications for UWB systems and hence there exist different operations for UWB systems. Herein, for illustration purposes, we consider a specific application of sensing a target represented by a stratified structure

C. Nguyen, J. Han, *Time-Domain Ultra-Wideband Radar, Sensor and Components*, SpringerBriefs in Electrical and Computer Engineering, DOI 10.1007/978-1-4614-9578-9_2, © Springer International Publishing Switzerland 2014

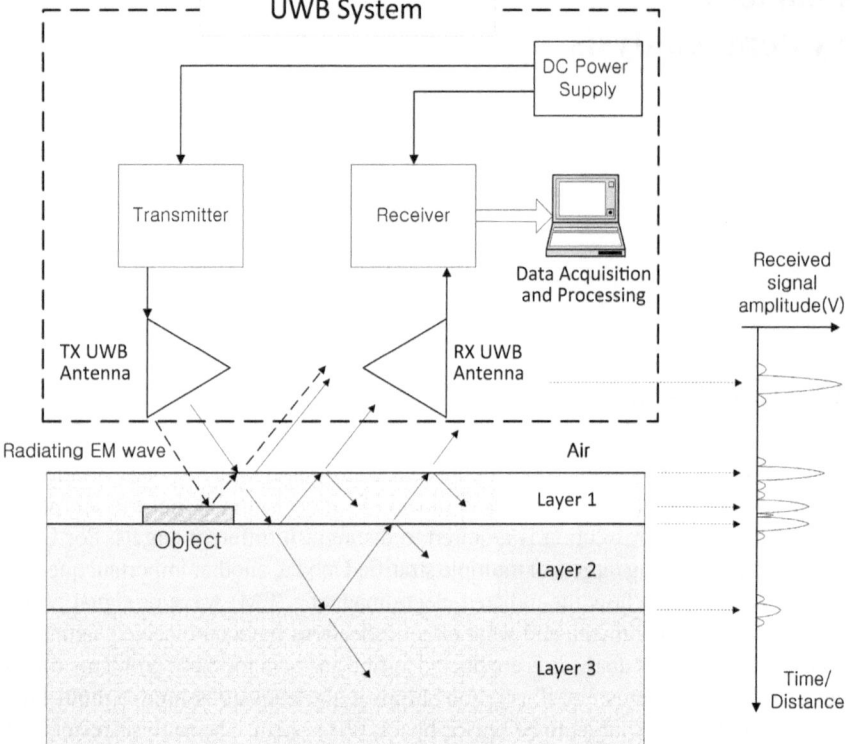

Fig. 2.1 UWB system and its operating principle in sensing the internal structure of a stratified-medium target

containing an object. Figure 2.1 shows a bistatic UWB system having separate transmitting and receiving antennas and its operational principle for this specific target. The UWB system is used to detect reflected signals from each interface between layers and the object. The first upper interface in Fig. 2.1 is formed between air and the first layer, and the second interface is formed between the first layer and the upper surface of the object, etc. The UWB system consists of a transmitter, a receiver, antennas, and a data acquisition and processing unit, similar to other sensing systems.

In Fig. 2.1, an EM pulse radiated from the transmitting antenna impinges on the surface of the target. Part of the incident wave to the surface is reflected back and captured by the receiving antenna. The remaining is transmitted into the first layer. This kind of reflection and transmission occurs on every layer interfaces as described in Fig. 2.1, and some of the reflected waves from each interface are captured by the receiving antenna. The received signal from the receiving antenna can be represented in the time domain as shown on the right-hand side of Fig. 2.1. Using this kind of received signals, we may identify the relative location of each interface and, eventually, the internal structure of the target. Further signal processing such as image processing can give more detection information for the layers and the object embedded within the target.

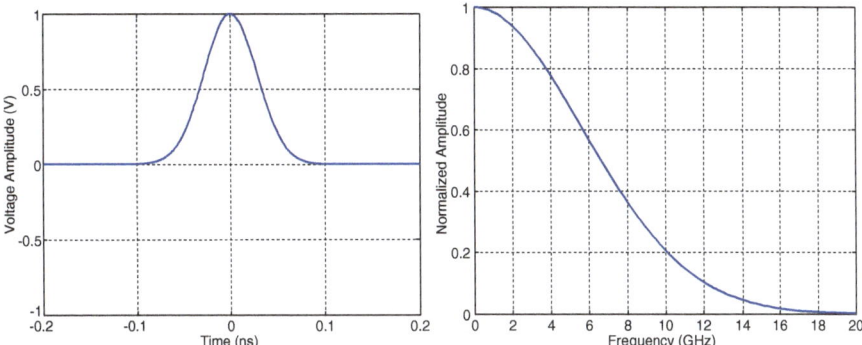

Fig. 2.2 Gaussian impulse with 200-ps pulse duration and its frequency spectrum

2.3 UWB Signals

The selection of impulse-signal types for UWB systems is one of the fundamental considerations in designing UWB systems, antennas, and circuits because the type of an impulse determines the UWB signal's spectrum characteristic. Many types of impulse signals such as step pulse, Gaussian-like (or monopolar) impulse, Gaussian-like single-cycle (or monocycle) pulse, Gaussian-like doublet pulse, and multi-cycle pulse can be used for UWB systems. Among those, Gaussian-like impulse, doublet pulse, and monocycle pulse are typically used in UWB systems. Particularly, the monocycle pulse is preferred in most UWB systems because of its spectral characteristics that facilitate easier wireless transmission than the impulse, wider bandwidth than the multi-cycle pulse, and easier to realize than the doublet pulse.

2.3.1 Gaussian Impulse

Figure 2.2 shows the time-domain waveform of a Gaussian impulse that has a shape of the Gaussian distribution, along with its frequency-domain waveform or spectral response. The impulse is assumed to have 200-ps pulse duration (or pulse width). The Gaussian impulse can be expressed as

$$y(t) = Ae^{-a^2t^2} \tag{2.1}$$

where A is the maximum amplitude of the Gaussian impulse and a is the constant that determines the slope of the Gaussian pulse. The spectral response containing the spectral components of the Gaussian impulse is obtained by taking its Fourier transform as

$$Y(\omega) = \frac{A}{a\sqrt{2}} e^{\frac{\omega^2}{4a^2}} \tag{2.2}$$

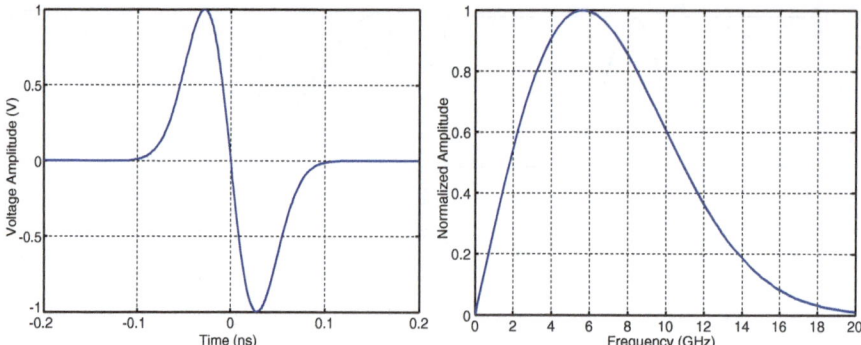

Fig. 2.3 Gaussian monocycle pulse with 200-ps pulse duration and its frequency spectrum

The frequency corresponding to the peak value of the impulse in the frequency domain is $f_o = 0$. The 3-dB bandwidth of the Gaussian impulse can be derived by letting the amplitude of the impulse at the 3-dB band-edge equal to the $1/\sqrt{2}$ of the maximum value at $f = 0$ as

$$\Delta f = 0.8326 \frac{a\sqrt{2}}{2\pi} \tag{2.3}$$

2.3.2 Gaussian Monocycle Pulse

Gaussian monocycle pulse is the first derivative of the Gaussian impulse signal. Figure 2.3 shows a Gaussian monocycle pulse having the same 200-ps pulse duration as the Gaussian impulse shown in Fig. 2.2 and its spectrum. The Gaussian monocycle pulse is described by

$$y(t) = -a^2 A t e^{-a^2 t^2} \tag{2.4}$$

The spectral response of the Gaussian monocycle pulse is given as

$$Y(\omega) = \frac{i\omega A}{a\sqrt{2}} e^{-\frac{\omega^2}{4a^2}} \tag{2.5}$$

The frequency corresponding with the peak value of the Gaussian monocycle pulse in the spectrum is obtained as

$$\Delta f_o = \frac{a\sqrt{2}}{2\pi} \tag{2.6}$$

and the 3-dB bandwidth can be derived as

Fig. 2.4 Gaussian monocycle pulses with different pulse durations

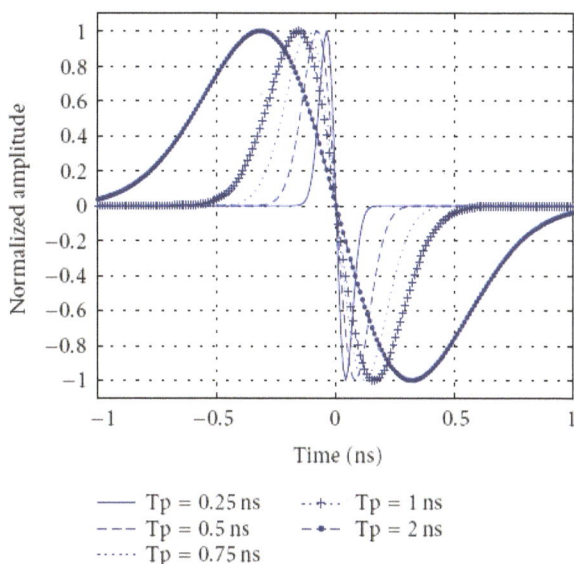

— Tp = 0.25 ns   ··+·· Tp = 1 ns
– – Tp = 0.5 ns   –•– Tp = 2 ns
····· Tp = 0.75 ns

$$\Delta f = 1.155\frac{a\sqrt{2}}{2\pi} = 1.155 f_o = \frac{1.155}{T_p} \tag{2.7}$$

where $T_p = 1/f_o$ is the pulse duration , which shows that the 3-dB bandwidth of the Gaussian monocycle pulse is approximately equal to 115 % of the pulse's center frequency f_o. Figs. 2.4 and 2.5 show the waveforms and spectrums of various Gaussian monocycle pulses having different pulse durations.

2.3.3   Gaussian Doublet Pulse

Figure 2.6 shows a Gaussian doublet pulse having 200-ps pulse duration and its spectrum. The Gaussian doublet pulse is the second derivative of the Gaussian impulse signal and hence can be expressed as

$$y(t) = -2a^2 A^{-a^2 t^2}(1 - 2a^2 t^2) \tag{2.8}$$

The spectral response of the Gaussian doublet pulse is

$$Y(\omega) = \frac{-A\omega^2}{a\sqrt{2}} e^{\frac{\omega^2}{4a^2}} \tag{2.9}$$

The frequency at which the peak value of the Gaussian doublet pulse occurs in the spectrum is

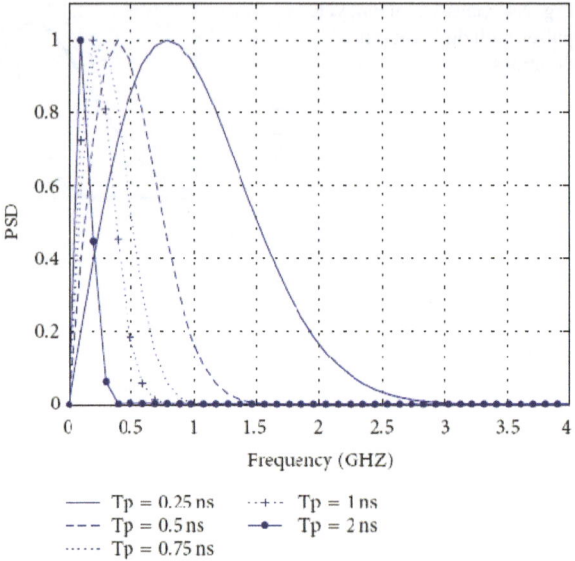

Fig. 2.5 Spectrum of Gaussian monocycle pulses with different pulse durations

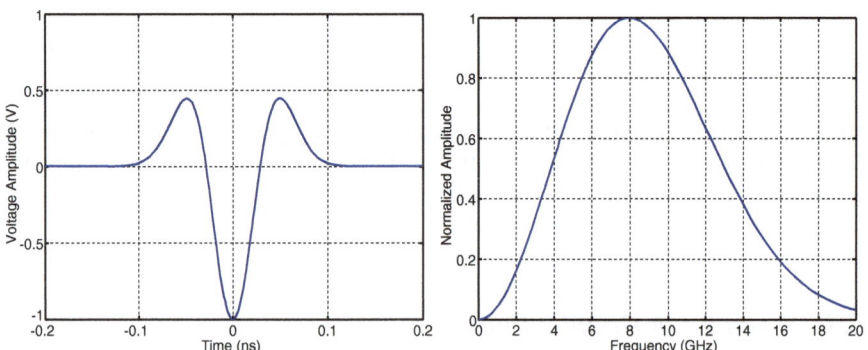

Fig. 2.6 Gaussian doublet pulse with 200-ps pulse duration and its frequency spectrum

$$f_o = \frac{a}{\pi} \tag{2.10}$$

This frequency is higher than that given in (2.6) for the Gaussian monocycle pulse. The 3-dB bandwidth can be derived as

$$\Delta f = 1.155 \frac{a\sqrt{2}}{2\pi} = 1.155 \frac{f_o}{\sqrt{2}} \tag{2.11}$$

Compared to the bandwidth of the Gaussian monocycle pulse given in (2.7), the absolute bandwidth of the Gaussian doublet pulse is same, yet the fractional bandwidth is larger assuming the same pulse duration. This result is due to the second-derivative performed upon the Gaussian impulse. Additional derivatives taken on the Gaussian impulse would produce other pulses having the same pulse duration but with progressively increasing fractional bandwidth and frequency corresponding to the peak pulse-magnitude. This phenomenon further implies that UWB signals generated using higher derivatives of the Gaussian impulse may be attractive for high-frequency UWB systems since they have higher frequencies and larger fractional bandwidth for the same pulse duration, which may be useful for some applications. It is noted that using a Gaussian monocycle pulse, which is the first derivative of a Gaussian impulse, at high frequencies requires a very narrow pulse duration which may be difficult to realize with sufficient amplitude in practice.

As can be seen from the pulse waveforms, the Gaussian impulse has no zero crossing point, while the Gaussian monocycle pulse and Gaussian doublet pulse have one and two zero crossings, respectively, which help define the bandwidth characteristics of these pulses. It is also observed that the spectral responses of these pulses contain no side-lobes beyond the zero-crossing frequency points which are desirable for signal transmission. For pulses whose spectral responses have side-lobes, such as a rectangular or sinusoidal pulse, these side-lobes are always outside the pass-band, which at most extends across the zero-crossing frequency ends, and hence produce unwanted radiation, leading to possible false-target detection and/or interference to other existing systems, especially when they have sufficiently high energy.

It is particularly noted that, as the peak spectral amplitude of the Gaussian impulse occurs at DC and as seen in Fig. 2.2, the bulk of its energy is contained at DC and low frequencies near DC. The monocycle and doublet pulse signals, on the other hand, contain no DC component and have much lower low-frequency energy. In general, the monocycle and doublet pulses have similar energy distributions in the low- and high-frequency regions around the center frequency. It is the difference in the spectral shapes of these signals at DC and low frequencies that greatly affects the transmission of signals via antennas and the propagation of signals though components, and ultimately the design of UWB antennas, components and systems. Impulses are not transmitted and received effectively through practical antennas due to their large portion of low-frequency spectral components which cannot be transmitted (or is transmitted with very low efficiency) by practical antennas. Monocycle and doublet pulses, on the other hand, can be transmitted more efficiently due to no DC component and less low-frequency content. Furthermore, using monocycle or doublet pulse facilitates the design of components including antenna in UWB systems due to no design consideration at DC and less design emphasis at low frequencies, leading to simpler and more compact design. It is further noted that signal fidelity is of utmost important for UWB systems which require signals to be transmitted and received with minimum distortion. With no DC component and less low-frequency spectral amplitudes contained in monocycle pulses, antennas and other system components can be more conveniently designed to cover desired bandwidth, hence minimizing the distortion of signals traveling through these components and, consequently, producing better fidelity for transmitting and receiving signals.

Fig. 2.7 Simplified diagram
of incident and reflected signals on a two-layer object

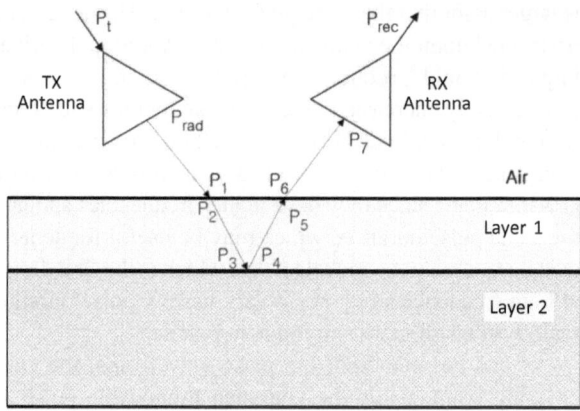

UWB systems always transmit a train of pulses (typically periodically) instead of
a single pulse. Consequently, according to Fourier analysis, the spectrums of UWB
pulse signals are not continuous and contain discrete spectral lines (corresponding to
discrete frequencies) spaced apart by $1/T$, where T is the period of the UWB signals.
Fourier analysis also shows that a UWB signal consisting of a train of pulses is not
substantially distorted by passive components including antennas having a bandwidth approximately equal to the reciprocal of the pulse width, because of most of
the energy is contained within such bandwidth. According to the Parseval's theorem, the average power in a periodic pulse train is equal to the sum of the powers in
its spectral components including DC and harmonics. Therefore, transmission of a
UWB signal consisting of periodic high-voltage pulses would be similar to simul-
taneous transmission of strong CW signals at different frequencies. The results of
the Parseval's theorem also suggest an alternate way of generating a UWB signal
of periodic pulses by combining various CW signals having appropriate amplitudes
and frequencies.

2.4   Power Budget Analysis

The power budget analysis involves estimating the minimum required transmitter
output power to produce detectable reflected signals from a target. To illustrate this
analysis, we consider a target consisting of two dielectric layers as shown in Fig. 2.7
together with the incident and reflected signals transmitted and received by the
transmitter (TX) and receiver (RX) antennas of a system, respectively. For simplic-
ity without loss of generality, we only consider the signal reflected from the second
interface. That is, we are only interested in determining the power budget involved
with the first layer and neglect the signals entering and reflected from the second
layer 2. The analysis can be easily extended for multiple layers.

In Fig. 2.7, the signals transmitted and received by the transmitter and receiver,
respectively, is represented by a power flow diagram that consists of the following

powers: P_t (transmitter's output power), P_{rad} (radiating power from the transmitting antenna), P_1 (incident power from air to interface 1 or air/layer 1 interface), P_2 (transmitting power from interface 1 into layer 1), P_3 (incident power from layer 1 to interface 2 or layer 1/layer 2 interface), P_4 (reflected power at interface 2), P_5 (incident power from layer 2 to interface 1), P_6 (transmitting power from interface 1 to air), P_7 (incident power from air into the receiving antenna), and P_{rec} (received power from the receiving antenna).

The received power P_{rec} in dBm can be expressed as

$$P_{rec}(dBm) = P_t + L_t \tag{2.12}$$

where L_t is the total loss defined as

$$L_t \equiv \frac{P_{rec}}{P_t} \tag{2.13}$$

The receiver sensitivity, S_i, in dB can be represented, using the required minimum transmitter's output power, $P_{t,\ min}$, as

$$S_i(dB) = P_{t,min} + L_t \tag{2.14}$$

which shows that the required minimum transmitter output power can be determined from the total loss and the receiver sensitivity determined by the receiver performance. For a given receiver, the main problem to determine $P_{t,\ min}$ is the calculation of the total loss. The total loss can be expressed using the powers defined in Fig. 2.7 as

$$L_t = \frac{P_{rec}}{P_7}\frac{P_7}{P_6}\frac{P_6}{P_5}\frac{P_5}{P_4}\frac{P_4}{P_3}\frac{P_3}{P_2}\frac{P_2}{P_1}\frac{P_1}{P_{rad}}\frac{P_{rad}}{P_t} \tag{2.15}$$

The power ratios in (2.15) can be grouped into several loss factors according to the cause of the loss [1]. We specify these loss factors as the antenna loss (L_{ant}), spreading loss (L_s), material attenuation loss (L_a), transmission coupling loss (L_{t1}), retransmission coupling loss (L_{t2}), and target scattering loss (L_{sc}). These loss factors are described as

$$L_{ant} = \frac{P_{rec}}{P_7}\frac{P_{rad}}{P_t} \tag{2.16}$$

$$L_s\ L_a = \frac{P_1}{P_{rad}}\frac{P_3}{P_2}\frac{P_5}{P_4}\frac{P_7}{P_6} \tag{2.17}$$

$$L_{t1} = \frac{P_2}{P_1} \tag{2.18}$$

$$L_{t2} = \frac{P_6}{P_5} \tag{2.19}$$

$$L_{sc} = \frac{P_4}{P_3} \qquad (2.20)$$

It is noted that the antenna loss (L_{ant}) represents the total loss incurred by both transmitting and receiving antennas. The loss of each antenna can be divided further into the antenna efficiency (L_e) and the antenna mismatch loss (L_m). For instance, the antenna loss due to the transmitting antenna can be represented as

$$\frac{P_{rad}}{P_t} = \left(\frac{P_{rad}}{P_a}\right)\left(\frac{P_a}{P_t}\right) = L_e \, L_m \qquad (2.21)$$

where P_a is the actual power entering the antenna. Using these definitions for loss factors, the total loss, L_t, in dB can be represented in terms of loss factors as

$$L_t(dB) = 2 L_e + 2 L_m + L_s + L_a + L_{t1} + L_{t2} + L_{sc} \qquad (2.22)$$

In order to calculate the total loss, a detailed model for each loss factor is needed. The antenna efficiency (L_e) and mismatch loss (L_m) can be assumed simply as -1 dB which is reasonable for well-designed antennas. The other loss factors are elaborated as follows.

Spreading Loss (L_s) Spreading loss occurs due to reduction of the power density of a wave with distance as it propagates. The following well-known radar equation represents the spreading loss in a general form for bistatic systems:

$$\frac{P_{ra}}{P_{ta}} = \frac{A_{et} \, A_{er} \, \sigma}{4\pi \, R^4 \lambda^2} \qquad (2.23)$$

where P_{ra} is the power received at the receiving antenna, P_{ta} is the power radiated from the transmitting antenna, A_{et} and A_{er} represent the effective apertures of the transmitting and receiving antennas, respectively, σ is the RCS (Radar Cross Section) of the target, R is the range from the system to the target, and λ is the operating wave length. The RCS is not considered in the spreading loss here; it will be dealt with in the target scattering loss later. After factoring out the RCS term, (2.23) can be simplified as

$$L_s = \frac{G_t \, A_{er}}{(4\pi R^2)^2} \qquad (2.24)$$

where G_t is the transmitting antenna gain. Equation (2.24) indicates that the spreading loss is a function of the inversed 4th power of the range. This relationship with the range is reasonable only for a point reflector type target. However, in the case of a planar reflector type such as ground interface, the spreading loss expression needs to be modified and can be approximated as

$$L_s = \left.\frac{P_7}{P_{rad}}\right|_{L_a=0dB} \cong \frac{G_t \, A_{er}}{(4\pi)^{3/2} \, R^2} = \frac{G^2 \lambda^2}{(4\pi)^{5/2} \, R^2} \qquad (2.25)$$

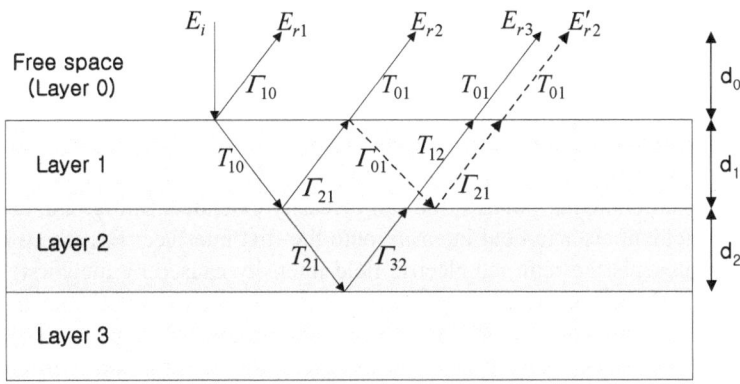

Fig. 2.8 Main reflections and transmissions in multi-layer structure

where G is the antenna gain assuming identical transmitting and receiving antennas. L_s is expressed in terms of the powers P_7 and P_{rad} defined in Fig. 2.7. In (2.25), no material attenuation effect is involved and the wavelength λ is not a single value due to wave propagation in different media such as those depicted in Fig. 2.7.

Material Attenuation Loss (La) EM waves propagating in a (practical) lossy medium experiences loss or attenuation of its power. The attenuation constant α of a lossy material can be found in the expression for the complex propagation constant γ of lossy materials as

$$\gamma = \alpha + j\beta = j\omega\sqrt{\mu\varepsilon'\left(1 - j\frac{\varepsilon''}{\varepsilon'}\right)} \tag{2.26}$$

where β is the phase constant, ε' and ε'' are the real and imaginary parts of the material's complex dielectric constant, μ is the permeability of the material, and ω is the radian frequency. The attenuation constant α of a material can be approximately obtained by an expansion of (2.26) as

$$\alpha = \omega\sqrt{\frac{\mu\varepsilon'}{2}(\sqrt{1 + \tan^2\delta} - 1)} \tag{2.27}$$

where $tan\delta \equiv \varepsilon''/\varepsilon'$ is the loss tangent of the material. The total attenuation or loss in dB for waves propagating a distance R in a material is then expressed as

$$L_a(dB) = -8.686\omega R\sqrt{\frac{\mu\varepsilon'}{2}(\sqrt{1 + \tan^2\delta} - 1)} \tag{2.28}$$

Transmission Coupling Loss (L_{t1}), Retransmission Coupling Loss (L_{t2}), and Target Scattering Loss (L_{sc}) Transmission coupling loss (L_{t1}), retransmission coupling loss (L_{t2}), and target scattering loss (L_{sc}) are related to the reflection and transmission of EM waves at the interface of different media. Figure 2.8 shows a target consisting of multiple layers of materials and a simplified reflection-transmission diagram illustrating multiple reflections and transmissions of electric fields (and hence waves) occurred in the layers.

In Fig. 2.8, for simple analysis, it is assumed that the radiated electric field from the transmit antenna impinges perpendicularly on the surface of the multi-layer structure. The reflections and transmissions for oblique incidence of waves with parallel or perpendicular polarization can be easily extended. In Fig. 2.8, E_i represents the incident electric field intensity onto the first interface; E_{rk}, where k is an integer, represents the returned electric field intensity caused by the (first) single reflection at the kth interface; E_{rk}' represents the returned electric field intensity caused by the (subsequent) double reflections at the kth interface, which is different from the reflection described by E_{rk}; T_{mn} represents the transmission coefficient for the transmitted signal from the nth layer to mth layer; Γ_{mn} represents the reflection coefficient for the incident signal from the nth layer to mth layer. It is noted that the reflection-transmission diagram depicted in Fig. 2.8 is simplified to include only the main reflection-transmission pairs that significantly contribute to producing a detectable returned signal; it does not include all possible reflection and transmission pairs that would occur in multi-layer structures.

Using the reflection-transmission diagram in Fig. 2.3, we can obtain the expressions for the returned electric field intensities with respect to the incident field intensity, E_i, as

$$\frac{E_{r1}}{E_i} = \Gamma_{10} \tag{2.29}$$

$$\frac{E_{r2}}{E_i} = T_{10}\Gamma_{21}\,T_{01} \tag{2.30}$$

$$\frac{E_{r3}}{E_i} = T_{10}\,T_{21}\Gamma_{32}\,T_{12}\,T_{01} \tag{2.31}$$

$$\frac{E_{r2}'}{E_i} = T_{10}\Gamma_{21}\Gamma_{01}\Gamma_{21}\,T_{01} = T_{10}\Gamma_{21}^2\Gamma_{01}\,T_{01} \tag{2.32}$$

It is noted that, the returned electric field intensity including double reflections at the 2nd interface, E_{r2}', expressed in (2.32), is much smaller than others described in (2.29)–(2.31) because, in most sensing applications, the magnitude of the reflection coefficient is much smaller than that of the transmission coefficient and E. (2.32) includes three times more of reflections than others. Consequently, the resulting magnitude of (2.32) is much smaller than others and hence can be ignored. Equations (2.29)–(2.31) involving a single reflection can be written in general as

$$\frac{E_{rn}}{E_i} = \Gamma_{n,n-1}\prod_{m=1}^{n-1}(T_{m,m-1}\,T_{m-1,m}) \tag{2.33}$$

The reflection and the transmission coefficients in (2.33) can be calculated for normal incident waves as

$$\Gamma_{n,n-1} = \frac{\eta_n - \eta_{n-1}}{\eta_n + \eta_{n-1}} \tag{2.34}$$

$$T_{n,n-1} = \frac{2\eta_n}{\eta_n + \eta_{n-1}} \tag{2.35}$$

where $\eta_n = \sqrt{\mu/\varepsilon_n}$ is the intrinsic impedance of the nth layer with μ and ε_n being the permeability and permittivity of the nth layer, respectively. The intrinsic impedance of lossy materials is complex; therefore, the reflection and transmission coefficients for lossy materials are also complex as well. To simplify the analysis, however, we assume low-loss materials and hence the intrinsic impedances can be assumed to be real.

The loss factors related to the reflection and transmission of signals are divided into two groups: one is L_{sc} relating to the reflection, and the other is L_{t1} and L_{t2} relating to the transmission. Since the reflected power from a target is related to the target's RCS as well as the reflection coefficient, the target scattering loss L_{sc} can be defined as [1]

$$L_{sc} = \frac{P_{ref}}{P_{inc}} = \sigma |\Gamma|^2 \tag{2.36}$$

or, in dB,

$$L_{sc}(dB) = 20\log|\Gamma| + 10\log\sigma \tag{2.37}$$

where P_{ref} is the reflected power , P_{inc} is the incident power, and σ is the RCS of the target. Since the RCS value of a dielectric half-space such as the ground is known as 1, it can be ignored in our analysis. The target scattering loss can therefore be approximated as a simple multiplication of all the reflection coefficients occurred at the interfaces on the signal propagation path. The transmission and retransmission coupling loss, L_{t1} and L_{t2}, are in general the multiplication of the transmission coefficients on two different propagation paths, one in a downward direction and the other in an upward direction, respectively. As a result, Eq. (2.33) turns out to be the total loss that includes all the loss factors related to the signal reflection and transmission effects on a single returned signal, E_{rn}. This new total loss factor is defined as the transmission loss L_u:

$$L_u(dB) = L_{sc} + L_{t1} + L_{t2} = 20\log\left|\frac{E_{rn}}{E_i}\right| \tag{2.38}$$

Required Minimum Transmitting Power

Calculations of the loss factors, according to the derived equations, require values of the physical and electrical parameters characterizing interested targets.

Table 2.1 Parameters of a typical pavement structure

Layer	Thickness range (inch)	Typical thickness (inch)
Asphalt	2–10	6
Base	4–14	10
Sub-base	N/A	8

Table 2.2 Electrical properties of a pavement structure. $\varepsilon_r'=\varepsilon_r$ and ε_r'' are the real and imaginary parts of the relative dielectric constant; α is the attenuation constant calculated from (2.27); and $\eta \cong 120\pi / \sqrt{\varepsilon_r}$ is the intrinsic impedance

Layer	ε_r'	ε_r''	α (Np/m) at 2 GHz	η (Ω)
Asphalt	5–7	0.03–0.05	0.22–0.47	142–168
Base	8–12	0.3–0.8	1.8–5.9	108–133
Sub-base	20	N/A	N/A	84

As an example, we list in Table 2.1 the physical parameters of a typical pavement structure consisting of (top) asphalt, base, and (bottom) sub-base with normal physical dimensions. Table 2.2 summarizes the electrical parameters or properties of each layer of the pavement structure.

The required minimum transmitting output power for the considered payment structure can now be calculated based on the parameters in Table 2.1 and 2.2. The transmitting output power should be sufficient to detect the most distant object which is the 3rd interface between the base and sub-base layers of the pavement structure in this example. Therefore, the calculation of the required output power should be based on the returned signal from the 3rd interface. Let us assume that the interested frequency is 2 GHz which is the center frequency of a transmitting monocycle pulse with 400-ps duration. The thickness of the asphalt and base layers, depicted as d_1 and d_2 in Fig. 2.8, is assumed as 6 and 10 in., respectively. The distance between the end of the antenna aperture and the top of the asphalt layer, depicted as d_0 in Fig. 2.8, is assumed to be 10 in.

First, the spreading loss L_s is calculated using (2.25). The same antenna is used for the transmitting and receiving antennas, and the antenna gain G is assumed to be 10 dB which is reasonable for well-designed antennas such as the microstrip quasi-horn antenna to be described in Chap. 5. The range R is 50 in., which is the sum of d_0, d_1 and d_2. The calculated L_s is -22.6 dB.

Second, the total material loss L_a is determined as the sum of the losses in all layers calculated using (2.28). Let L_{a1} and L_{a2} be the loss per unit length in dB/m for the asphalt (layer 1) and base (layer 2), respectively. These losses can be obtained as -8.686α, where α is the attenuation constant of the individual layer in Neper/m. Calculation results using the maximum values for α provided in Table 2.2 give $L_{a1}=-4.08$ dB/m and $L_{a2}=-51.2$ dB/m. The traveling distance in each layer is twice of its thickness ($2d_1=0.5$ m and $2d_2=0.7$ m). Therefore, the total material loss in dB is $L_a=2L_{a1}d_1+2L_{a2}d_2=-37.9$ dB.

Third, the transmission loss L_u defined as (2.38) is calculated. As mentioned earlier, we are interested in the detection of the farthest object which is the 3rd interface, which necessitates the calculation of $L_u = 20\log|E_{r3}/E_i|$ using (2.31). The reflection and the transmission coefficients included in (2.31) can be calculated using (2.34) and (2.35), respectively. The intrinsic impedances for the layers are obtained as $\eta_0 = 377\ \Omega$, $\eta_1 = 150\ \Omega$, $\eta_2 = 120\ \Omega$, and $\eta_3 = 84\ \Omega$ based on the electrical properties of these layers listed in Table 2.2. The transmission loss can now be calculated as $L_u = -17$ dB.

Finally, using the calculated loss factors, the total loss can be obtained as $L_t = 2$ $(L_e + L_m) + L_s + L_a + L_u = -82$ dB, where the antenna efficiency L_e and antenna mismatch loss L_m are assumed as -1 dB.

To obtain the required minimum transmitter output power according to (2.14), we need to know the receiver sensitivity S_i besides the total loss. Assume 8-dB tangential sensitivity for the receiver, which is generally used for measurement systems [2], we can express the receiver sensitivity in dB as

$$S_i(dB) = kTBF + SNR_o = kTBF + 8 \tag{2.39}$$

where $k = 1.38 \times 10^{-23}$ JK^{-1} is the Boltzmann's constant, T is the absolute temperature in Kelvin degree (°K), B is the receiver's bandwidth, F is the receiver's noise figure, and SNR$_o$ is the required output SNR (Signal-to-Noise Ratio), which is 8 dB for the 8-dB tangential sensitivity. The calculation result for the receiver sensitivity using T = 298 °KK (room temperature), B = 5 GHz and F = 3 dB is $S_i = -66$ dB. Using (2.14) and the calculation result for the total loss and sensitivity, the required minimum transmitter power is obtained as $P_{t,\ min} = 16$ dBm. Note that this is the average power for the CW signal, which is at 2 GHz as considered here. For a UWB signal such as the monocycle pulse, we are interested in the pulse peak power or peak-to-peak voltage value (V_{pp}). The conversion result from the average power to the peak-to-peak voltage value is 4 V_{pp} for 50-Ω load, which is the required minimum voltage of the transmitter output pulse and is equivalent to 80 mW of the peak power.

2.5   Range Resolution Analysis

The minimum range resolution for UWB systems is required only for the detection of the thinnest layer in a multi-layer target. In our considered example of the pavement structure, the thinnest layer is the asphalt layer. Let us assume that the minimum thickness of the asphalt we need to discern from the base layer is 1 in. Therefore, the objective of the resolution analysis is finding the minimum required pulse duration to achieve the required range resolution of 1 in. We can find the minimum required pulse duration T_p from the simple equation of $T_p = d_m/v_p$, where d_m is the minimum thickness of the asphalt layer and v_p is the phase velocity of the propagating wave at a certain frequency. For low-loss materials, $v_p \cong \omega/\beta$ for plane waves, where the phase constant β may be approximated for low-loss materials as

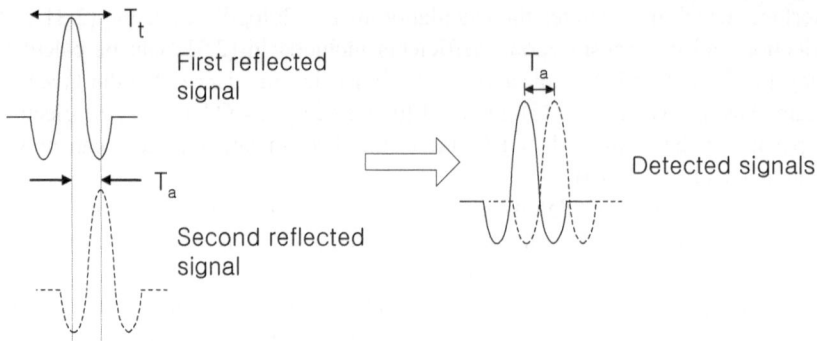

Fig. 2.9 Two detected signals with minimum discernable time interval

$$\beta \cong \omega\sqrt{\mu\varepsilon'}\left(1+\frac{1}{8}\tan^2\delta\right) \cong \omega\sqrt{\mu\varepsilon'} \tag{2.40}$$

where $\varepsilon' = \varepsilon_o\varepsilon_r$ with ε_o being the permittivity of air. The phase velocity v_p can then be approximated for low-loss non-magnetic materials as

$$v_P \cong \frac{1}{\sqrt{\mu\varepsilon'}} = \frac{1}{\sqrt{\mu_o\varepsilon_o\varepsilon_r}} = \frac{c}{\sqrt{\varepsilon_r}} \tag{2.41}$$

where $c = 3 \times 10^8$ m/s. The phase velocity in the asphalt layer, whose relative dielectric constant is given in Table 2.2, is calculated from (2.41) as 4.46×10^9 to 5.28×10^9 in./sec. The minimum required pulse duration for the transmitting pulse is then determined as $T_p \cong 200$ ps. This short pulse duration, however, does not take into account the actual waveform shape in the detection stage. If the actual received waveform from the receiving antenna is considered, then the minimum pulse duration needed for the required range resolution is not the same as the transmitting pulse duration.

As we will use the microstrip quasi-horn antennas described in Chap. 5 for the UWB system presented in this book, let's assume TEM horn antenna is used for the transmitting and the receiving antennas in this analysis. For this type of antenna, the radiating signal is the first derivative of the input signal [3–5]. Therefore, for an input signal of monocycle pulse, the received waveform through the receiving antenna is approximately similar to the Sinc-function as seen in Fig. 2.9. Figure 2.9 shows two reflected signals from two different layer interfaces separated by a distance equal to the minimum range resolution, which are detected in sequence, generated from a monocycle pulse transmitted by a TEM horn antenna. The first and second reflected signals are from the first and second interfaces, respectively. The pulse duration T_t is assumed to be the same as that of the transmitting pulse, ignoring typically small pulse-stretching effect of a well-designed TEM horn antenna and other pulse effects due to the reflection coefficients at the interfaces. As shown in the waveforms of the detected signals in Fig. 2.9, a minimum time interval T_a is required to completely discern the two detected signals without any overlap in the pulse main-lobes. In order to achieve that, T_a should be equal to the minimum required pulse duration, T_p. However, as shown in Fig 2.9, T_a is about a half of the transmitting pulse duration T_t. Therefore, the transmitting pulse duration T_t needed for achieving a range resolution of 1 in. is about twice the time of T_p, which is about 400-ps.

2.6 Summary

This chapter covers the theory and analysis of UWB systems, particularly the system operating principle, power budget and range resolution. Various UWB pulse signals commonly used for UWB systems, including Gaussian-like impulse, doublet pulse, and monocycle pulse, are addressed. Detailed calculations for the minimum requited transmitting power and minimum required pulse duration for a specific range resolution are also presented using a typical multi-layer pavement structure.

References

1. Daniels, D.J.: Surface Penetrating Radar. IEE Press, London (1996)
2. Fontana, R.J., Richley, E.A., Beard, L.C., Barney, J.: A programmable ultra wideband signal generator for electromagnetic susceptibility testing. In 2003 IEEE Conference on Ultra Wideband Systems and Technologies, pp. 21–25 (2003)
3. Taylor, J.D.: Introduction to Ultra-Wideband Radar Systems. CRC Press: Boca Raton (1995)
4. Theodorou, E.A., Gorman, M.R., Rigg, P.R., Kong, F.N.: Broadband pulse-optimised antenna. IEE Proceedings, vol. 128, pt. H, no. 3, pp. 124–130 (June 1981)
5. Miao, M., Nguyen, C.: On the development of an integrated CMOS-based UWB tunable–pulse transmit module. IEEE Transactions on Microwave Theory and Technique, vol. MTT-54, No. 10, pp. 3681–3687 (October 2006)

Chapter 3
UWB Transmitter Design

3.1 Introduction

The transmitter of UWB systems is basically a pulse generator which generates an impulse type signal and hence is much simpler than the transmitter used in CW systems. The pulse generator should have proper performance to satisfy required specifications for UWB system applications. The critical performance factors of the pulse generator as an UWB transmitter are the duration of the output pulse and its peak power. The duration of the output pulse should be small enough to satisfy the required resolution for detection of close targets. The peak power of the output pulse needs to be high enough to enable detection of long-range targets.

The output pulse of the pulse generator should also have good and symmetrical shape with small distortion in the main pulse and small ringing in the tail (or side-lobe) of the main pulse. An additionally desirable property for pulse generators is the tuning capability to generate different pulse durations with constant peak power and pulse shape. Tunable pulse generator can provide great flexibility in the operation of UWB systems with enhanced performance adaptable to different environments and targets. Pulse with wide duration contains large low-frequency components, enabling the pulse signal to propagate farther because of the relatively low propagation loss of its low-frequency components. Pulse with shorter duration, on the other hand, has wider frequency bandwidth, making feasible higher range resolution. A pulse that can change its duration, especially by an electronic means, would therefore have both advantages of increased range (or penetration) and fine range resolution, and is attractive for UWB systems, especially those intended for sensing of wide variety of targets that need varying ranges and resolutions. The pulse tuneability also provides the highly desired diversity for UWB transmitters (and hence systems), enabling them to function across multi-band which leads to enhanced information of targets. Electronically tunable pulse generators are also desired for measurement equipment. The polarimetric video impulse radar described in [1] is a good example showing the usefulness of the pulse generator's tuning capability.

Many kinds of pulse generators have been developed and used for UWB system applications. The pulse generators of UWB systems are distinguished by their

C. Nguyen, J. Han, *Time-Domain Ultra-Wideband Radar, Sensor and Components*,
SpringerBriefs in Electrical and Computer Engineering,
DOI 10.1007/978-1-4614-9578-9_3, © Springer International Publishing Switzerland 2014

output pulse peak power, pulse duration, and pulse shape. According to the peak power of the output pulse, the pulse generator can be divided into high-power and low-power pulse generators. A reference value of the peak power that can differentiate the high- and low-power pulse generators may be 500 mW which corresponds to 10 V_{p-p} in peak voltage amplitude of the pulse. The pulse generator for the UWB system design presented in this book, based on the required specifications of the output pulse peak power and its duration mentioned in Chap. 2, is categorized as low-power sub-nanosecond monocycle pulse generator. According to the pulse duration, the pulse generators can be divided into nanosecond and sub-nanosecond pulse generators. For instance, the output pulse duration of the sub-nanosecond pulse generators is smaller than 1 ns, which corresponds to bandwidth wider than 1 GHz approximately. According to the pulse shape, the pulse generator can be primarily classified as step function, impulse, monocycle-pulse, multi-cycle-pulse or doublet-pulse generator as addressed in Chap. 2.

In most cases of high-power applications for long-range detection, high-power pulse generators producing impulse signals of nanosecond duration are commonly used. Even though the impulse signal is not efficient for antenna transmission, as discussed in Chap. 2, it is widely used for these applications because it is easier and simpler to design high-power impulse generators than monocycle pulse generators. For high-power UWB applications, such as long-range UWB radar and deep-penetration ground penetrating radar (GPR), pulse generators using avalanche transistors with or without a series stack of step-recovery diodes (SRD) are commonly used [2, 3]. Optoelectronic design using GaAs photoconductive semiconductor switch and electronic design using drift step-recovery diode (DSRD) and silicon avalanche shaper are also well-known method for high-power pulse applications [4, 5]. These high-power pulse generators, however, have bulky, complicated structures and are expensive, and in some cases requires high DC bias voltage as much as 100 V. Devices such as DSRD are also difficult to find in commercial market. Commercially available high-power pulse generator have limited performance. Their achievable minimum pulse width is only about 1 ns and the maximum output pulse repetition frequency (PRF) is usually about 1 MHz.

Research in efficient pulse generator design for low-power, short-range UWB systems has not attracted much attention, partly because of the availability of commercial products with good performance [6, 7]. However, with the increasing demand of low-power and compact UWB systems, it is naturally desirable to have compact low-cost pulse generators. It is hard to find tunable pulse generators, especially those enabling seamless integration with antennas and being implemented with low cost, in commercial pulse generators. The pulse generators proposed in [8, 9] provide relatively high output pulse power and handy tuning capability, but the output waveform is an impulse or a step function. The tunable pulse generators reported in [10] also produce impulses only. The monocycle pulse generators developed in [11–15] provide relatively low output power (10 mW of peak power) without tuning capability. Several impulse and monocycle pulse generators with tunable durations were also developed [16–18].

In this chapter, we present the design of a sub-nanosecond tunable monocycle pulse transmitter for low-power short-range UWB applications. The developed monocycle pulse generator produces monocycle pulses with 10 MHz of PRF, tuning range of 0.4–1.2 ns for the pulse duration, corresponding approximately to 0.15–3.7 GHz operating frequency range. The output pulse amplitude is in the range of 6–9 V_{p-p} (peak-to-peak voltage), equivalent to 200–400 mW of peak power, which is comparable to the output power of commercial pulse products.

The developed tunable monocycle pulse transmitter makes use of a single SRD to form a fast transient rise-time, which is a commonly used approach for sub-nanosecond pulse generator design. Relatively high output pulse power is obtained by using high driving power for the SRD, provided by a driving circuit consisting of high-speed amplifier and buffer integrated circuits (ICs). This design approach greatly simplifies the circuit design as compared to that using discrete components proposed in [19] and avoid using expensive high-power wideband monolithic microwave integrated circuit (MMIC) at the output stage of the pulse generator as proposed in [13] and [14]. In addition to using high driving power, high output power can be obtained by increasing the output power efficiency of the SRD pulse generator using a proper DC biasing scheme for the SRD as described in [8, 19] and [20]. In the designed transmitter, a coupling circuit for the SRD is designed to increase the output power efficiency. This circuit has the form of a first-order high-pass RC filter and does not require any external DC bias for normal operation. A decoupling circuit is also employed to reduce the ringing level of the output pulse.

The basic idea for realizing the monocycle pulse tuning ability relies on using p-i-n-diode (or MESFET) switches proposed for tunable impulses [16]. The same tuning principle is extended to make a tunable monocycle rather than an impulse. In comparison with the tuning method using DC bias control for the SRD [8], the p-i-n diode (or MESFET) switching method is simpler, facilitates the design of tunable monocycle pulse waveforms and provides convenient tuning operation. In the designed monocycle pulse transmitter, an impulse is generated first and a pulse-shaping circuit is used to convert the impulse to a monocycle.

3.2   Design of Delay-Line SRD Impulse Generator

The basic element in the developed tunable monocycle pulse transmitter is the delay-line impulse generator. Formation of an impulse using delay line is a classical technique used in the digital or pulse circuit areas. Some SRD impulse or monocycle-pulse generators have been designed using the same delay-line principle and implemented on microstrip or coplanar waveguide (CPW) structures [11–14, 21]. Another type of SRD impulse generator, extensively used for sub-nanosecond pulse generation and microwave multiplier, is the shunt-mode SRD impulse generator [22–25].

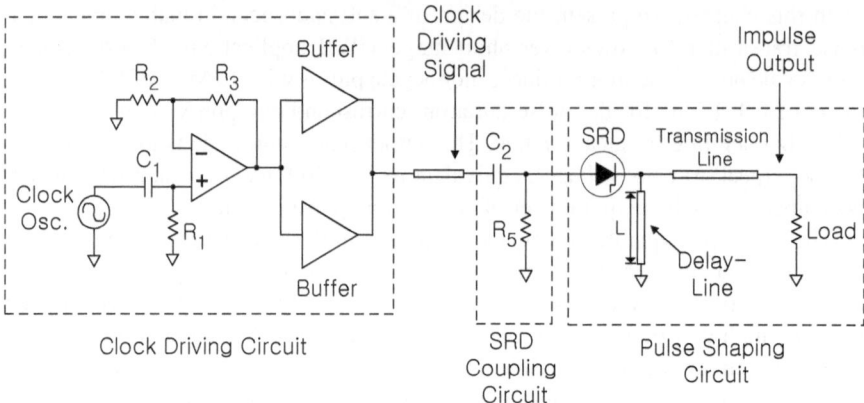

Fig. 3.1 Circuit diagram for the designed SRD delay-line impulse generator

Before beginning the circuit design, important SRD characteristics affecting the circuit performance of the impulse generator need to be identified. One of the important characteristics is the transition time of SRD that determines the minimum achievable pulse's transition time. The other is the minority carrier lifetime (MCLT) of the SRD, which affects the storage time under reverse bias conditions. If a clock signal is applied to the SRD pulse-shaping circuit, the rise time of the clock should be less than the MCLT of the SRD in order to obtain maximum achievable pulse amplitude [20]. In our design, the SRD is selected for 70-ps transition time and 10-ns MCLT, assuming that the output pulse duration to be generated is 200 ps and the rise time of the clock is 10 ns.

Figure 3.1 shows the overall circuit diagram of the designed impulse generator. It can be subdivided into the clock driving, the SRD coupling and the pulse shaping circuit according to their functions. The clock driving circuit consists of a clock oscillator, a non-inverted opamp voltage amplifier and two current boosting buffers. The clock oscillator generates a TTL (transistor-transistor logic) compatible clock signal with 5 ns of rise time and about 10 MHz of PRF. This fast rise-time clock signal is required according to the MCLT of the selected SRD. High voltage of the clock driving signal for the SRD, indicated in Fig. 3.1, produces large amplitude for the output impulse. To increase the driving signal voltage level, the input clock signal is amplified by amplifier and buffers. The desired voltage level of the driving signal is set as 20 V_{p-p}, taking into account the selected SRD having 15 V of breakdown voltage. The amplifier is designed to have a voltage gain of 4 in order to obtain 20 V_{p-p} from a 5 V_{p-p} TTL input signal and a bandwidth of 100 MHz using a wide-band opamp with 400-MHz unity-gain bandwidth in order to prevent the input rise time from slowing down. The buffer is used to supply the SRD, representing a low-impedance load, with enough current to obtain a high output voltage swing of 20 V_{p-p} and to reduce large loading effect on the amplifier. Two buffers are used to enhance the reduction of loading effect and to have some margin for its supplied current capacity. The loading circuit analysis, which will be shown below, determines the required current capacity for the buffers.

To produce an impulse having high amplitude and a transition as fast as possible, a proper DC bias can be applied to the SRD. The DC bias controls the quantity of the stored charge in the SRD through the forward-bias current, which determines the storage time and thus affecting the turn-off transition time when the SRD is reverse biased. The stored charge should be large enough to have a sufficient storage time so that a fast transition may occur at the maximum voltage level of the input clock signal. On the other hand, it should also be minimized to obtain as fast transition time as possible. In our design, we employ a SRD coupling circuit, as shown in Fig. 3.1, to avoid the use of an external bias network. The coupling circuit works effectively to control the quantity of the stored charge in the SRD. Detailed analysis of this circuit included in the following will explain this controlling effect and determine its optimal component values (R_s and C_2 in Fig. 3.1).

An equivalent circuit for the delay-line impulse generator is shown in Fig. 3.2a, in which the driving circuit in Fig. 3.1 is replaced by a voltage source with a source resistance. R_f represents the forward-biased resistance of the SRD. Figure 3.2b displays the voltages and currents in Fig. 3.2a. The source signal V_s is assumed as an ideal clock with instant transitions to simplify the analysis. The waveforms for V_c and V_r, defined in Fig. 3.2a, are drawn by means of a basic transient analysis. In the waveform of $V_r(t)$ (input to the pulse-shaping circuit), a fast transition occurred due to the turn-off of the SRD is illustrated. This transient is transformed to an impulse at the output by combining with the reflected one from the short-circuited transmission line representing the delay-line. The duration of the impulse is determined by the length of the delay-line. The waveform of the forward-biased current flowing through the SRD, which determines the stored charge quantity, is also shown with a peak value of I_m.

Our design goal is to determine optimum values for the resistors (R's) and capacitors (C's) in the equivalent circuit to produce a maximum transient amplitude of V_t for $V_r(t)$, described in Fig. 3.2b. The values for R's are determined qualitatively first as follows. To obtain sufficient storage time, the peak amplitude of the forward current, I_m, should be as large as possible. Since the forward current is proportional to the diode voltage,

$$I_m \propto V_m \propto -V_c(T^-) + V_s(T^+) = -V_x + V_{cf} \tag{3.1}$$

where V_{cf} is the peak voltage of the clock source that is 10 V in our case. From the relationship given in (3.1), in order to increase I_m, $|V_x|$ should be equal to V_{cf}. To increase $|V_x|$ and V_t, $\tau_{d2} = (R + R_s)C$ and $\tau_{d1} = (R_f + R_s)C$ should be decreased. Therefore, R, R_s and R_f should be as small as possible. However, R needs to be much greater than R_f to achieve a smooth transition after the fast diode turn-off transition in V_r. As a result, assuming R_f is equal to the series resistance of the SRD, which is 0.22 Ohm in our case, R = 10 Ω can then be selected as a reasonable value.

An optimal value for C can be found by deriving equations for the waveforms shown in Fig. 3.2b and solving them using an iterative optimization method. First, a

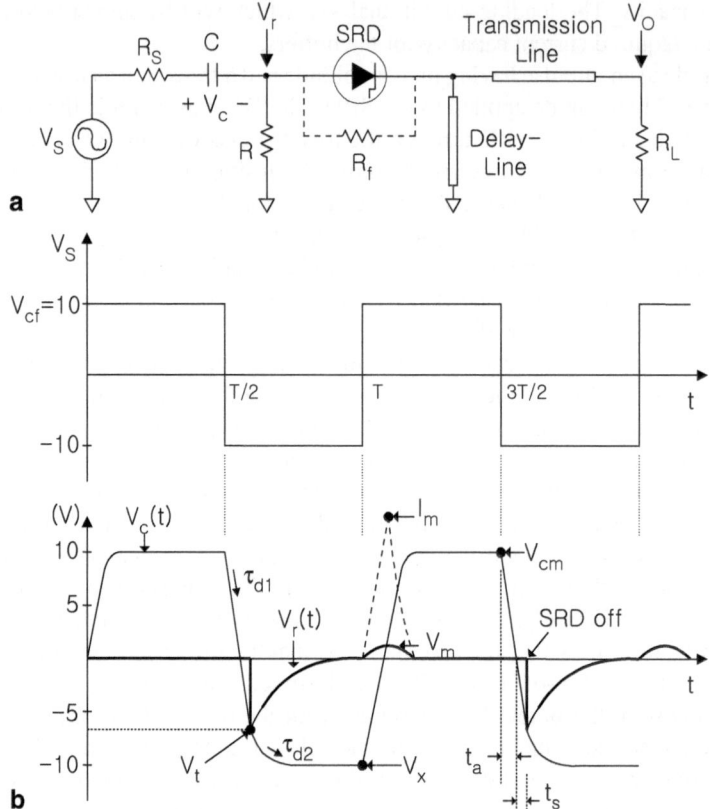

Fig. 3.2 Equivalent circuit of the SRD delay-line impulse generator **a** and the voltage and current waveforms occurred in the circuit **b**

small negative value for V_x is assumed as an initial value. From the rising and falling edges of the waveform $V_c(t)$, the following equations can be derived:

$$V_{cm} = V_{cf} + (V_x - V_{cf})e^{-T/2R_fC} \tag{3.2}$$

$$V_t = -V_{cf} + (V_{cm} + V_{cf})e^{-t_t/R_fC} \tag{3.3}$$

where $t_t = t_s + t_a$, with t_s and t_a being defined in Fig. 3.2b, and T represents the clock period. The storage time t_s can be calculated from the fact that the stored charge in the SRD resulted from the forward current flowing is the same as the removed charge caused by the reverse current. Assuming the forward current is proportional to applied diode voltage, it's straightforward to derive the following equation for the stored charge Q_f:

$$Q_f \cong \frac{V_a(t_k)}{R_f} \left[\frac{t_k}{2} + R_f C \left[e^{-t_k/R_f C} - e^{-t_L/R_f C} \right] \right] \tag{3.4}$$

where t_k is a half of the rise time t_r of the clock source, t_L is the MCLT of the SRD, and $V_a(t)$ is the response of a ramp signal with a slew rate of m, .

$$V_a(t) = m R_f C (1 - e^{-t/R_f C}) \tag{3.5}$$

with $m = |V_x|/t_k$. Similarly, the removed charge quantity Q_r can be derived for two distinct cases. For $t_s \le t_r$,

$$Q_r \cong \frac{V_a(t_s)}{2R_f} t_s \tag{3.6}$$

where $m = |V_{cm}|/t_k$. For $t_s > t_r$,

$$Q_r \cong \frac{V_a(t_k)}{R_f} \left[\frac{t_k}{2} + R_f C \left[e^{-t_k/R_f C} - e^{-t_s/R_f C} \right] \right] \tag{3.7}$$

The estimate of t_s can now be obtained by an iterative optimization method using the following criterion:

$$\hat{t}_s = \arg \left(\min_{t_0 \le t \le t_L} |Q_f - Q_r| \right) \tag{3.8}$$

where t_0 is a small time step used in the iteration process. From (3.3), with the condition of $V_t = 0$,

$$\hat{t}_a = -R_f C \left[\ln \left(\frac{V_{cf}}{V_{cm} + V_{cf}} \right) \right] \tag{3.9}$$

Using a negative part of the waveform $V_c(t)$ corresponding to the portion after the SRD turn-off transition, the updated estimate value of V_x can be calculated as

$$\hat{V}_x = -V_{cf} + (V_t + V_{cf}) e^{-(\frac{T}{2} - t_t)/RC} \tag{3.10}$$

Finally, the estimate of V_t can be obtained using the following criterion:

$$\hat{V}_t = \arg \left(\min_{n \to \infty} |\hat{V}_x(n) - \hat{V}_x(n-1)| \right) \tag{3.11}$$

where n represents the number of iterations in the optimization procedure. Results from the simulation and measurement are given in Fig. 3.3. Figure 3.3a shows

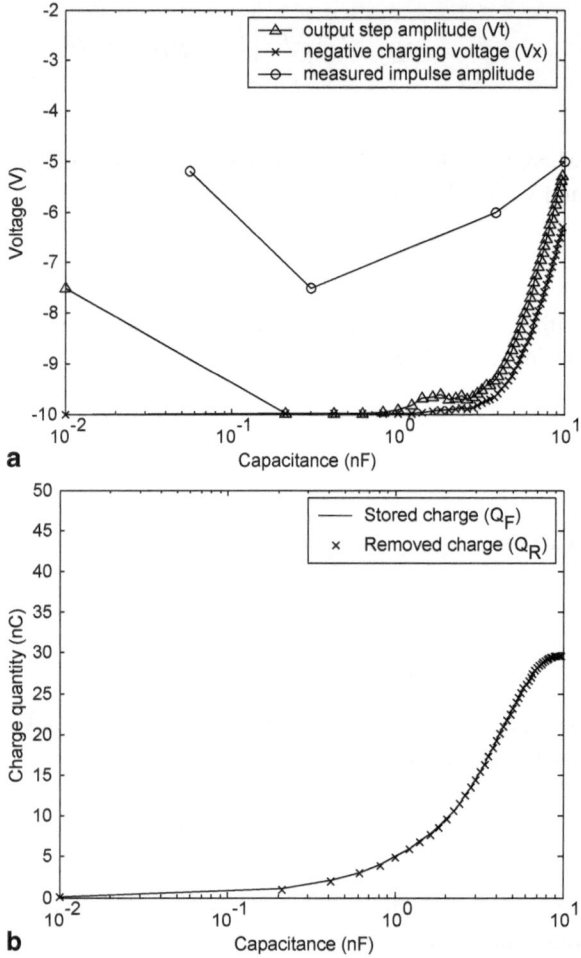

Fig. 3.3 a Calculated and measured voltages for the designed delay-line impulse generator. **b** Calculated results for the stored and removed charge quantities in the SRD

calculated values for the negative charging voltage V_x and the transient step voltage V_t along with the measured impulse amplitude as a function of the capacitance values of the RC coupling circuit. The simulation result for the step amplitude V_t shows a consistent trend with the measured impulse amplitude, thus validating the simulation. From this result, an optimal value for C can be chosen within the range of 0.2~1 nF. The simulation results for the stored and removed charge quantities, shown in Fig. 3.3b, confirms the fact that $Q_F \approx Q_R$, and hence verifies the convergence of the simulation. C = 200 pF is chosen as an optimal value based on the result in Fig. 3.3b considering the fact that smaller stored charge makes faster transition.

The required current capacity for the current boosting buffer can be calculated approximately based on the previous analysis. The loading circuit for the buffer is the same as that shown in Fig. 3.2a and the voltage waveform applied to the load is the same as $V_r(t)$ in Fig. 3.2b. Assuming R = 12 Ω, C = 1 nF, and $R_s + R_f = 1$ Ω, the

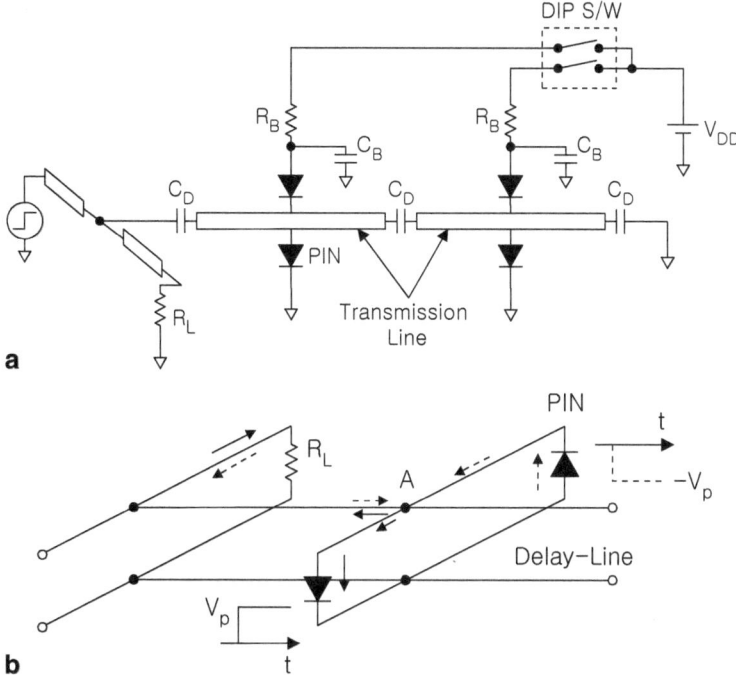

Fig. 3.4 **a** Circuit diagram for a distributed delay-line using anti-parallel p-i-n diode pairs. **b** Illustration showing the principle of using a p-i-n-diode pair to obtain a synthetic short-circuit. The *solid-* and *dashed-line arrows* represent current flows corresponding to the positive and negative step voltage transients, respectively

peak voltage amplitude with respect to the forward current is calculated as 2.77 V using (3.5). The required positive average current can then be calculated as 173 mA. Similarly, the negative average current required is determined as 210 mA. This result shows that the two parallel-connected current buffers, each with ±250 mA of output current capacity, can satisfy the driving current requirements with some design margin.

3.3   Design of Tunable Monocycle Pulse Generator

The tuning method used in our design is based upon the p-i-n diode switching method described in [16] for impulse generation. Here, it is extended to generate various monocycle pulse durations. The underlying principle of the tuning method using p-i-n diode switching is relatively simple. Figure 3.4a shows an instance of the distributed delay-line for tuning of the impulse duration. The original delay-line, seen in Figs. 3.1 and 3.2a, is subdivided into several transmission-line sections, separated by DC blocking capacitors, to form a distributed delay-line with each section containing an anti-parallel p-i-n-diode pair and a biasing circuitry. The length of each transmission line and the number of the transmission-line sections are chosen according to the desired duration and number of the impulses. The p-i-n diodes are forward- and reverse-biased to achieve on and off states, respectively, and each diode-pair is biased independently with each other. In operation, one diode-pair is turned on while all others are turned off simultaneously. A synthetic short-circuit can be created at each diode-pair connection point by turning *on* the p-i-n diodes through a DIP switch. By changing the DIP switch connections alternately, various delay-lines of different lengths can be effectively made, hence generating different impulse durations corresponding to different round-trip times of the step-function signal propagating on the delay-line. The DC blocking capacitor C_D is used in each p-i-n-diode/transmission-line section to provide DC-voltage decoupling between adjacent sections, whereas allowing the step-function signal to pass through. Use of a pair of anti-parallel p-i-n diodes, instead of a single diode, at each junction along the delay-line is explained in Fig. 3.4b. Let's assume a positive step pulse arrives at junction A on a delay-line section and both the p-i-n diodes are turned on. Assuming further that the step pulse is a large signal, then only two anti-parallel connected p-i-n-diodes can support opposite current directions required by the incident and reflected step pulses. An anti-parallel p-i-n diode-pair configuration, therefore, creates a synthetic short-circuit closer to the ideal one than a single-diode configuration for large signal input. The other advantage of using a pair of p-i-n diodes is that one of the diodes can be placed on the side of the bias circuit to reduce the coupling of the incident pulse to the bias circuitry, resulting in better isolation than using a large resistance for biasing and isolation purposes. Moreover, using two diodes supports the balance of the circuit configuration.

Figure 3.5 shows a simplified overall circuit diagram for the designed tunable monocycle pulse generator. The clock-driving and the SRD coupling circuits have the same configurations as those for the impulse generator described in Sect. 3.2. A broadband RF choke (RFC) is used to provide a return for the low-frequency clock driving signal to the ground, which is necessary in order to create a similar loading circuit as in the impulse generator design, so that the previous design result for the driving and coupling circuits of the impulse generator may be applied directly to the monocycle pulse generator. The RFC used in our design is ADCH-80A manufactured by Mini-Circuits Co. Two distributed delay-lines with identical configuration are used for tuning of the monocycle pulse duration. The first one is used for formation of various impulse durations. The other is used for controlling the round-trip time of the generated impulse, which results in monocycle pulses with different durations at the output load.

Direct coupling of the two distributed delay-lines may cause problem of backward transmission of the reflected impulse from the second delay-line to the first delay-line, which makes large multiple reflections and eventually causes large ringing on the output signal. A backward decoupling circuit is therefore needed between the two delay-lines to reduce the backward coupling effect. This decoupling circuit included in Fig. 3.5, consisting of a capacitor, resistor and Schottky diode, facilitates direct coupling of the incident pulse from the first delay-line into the second

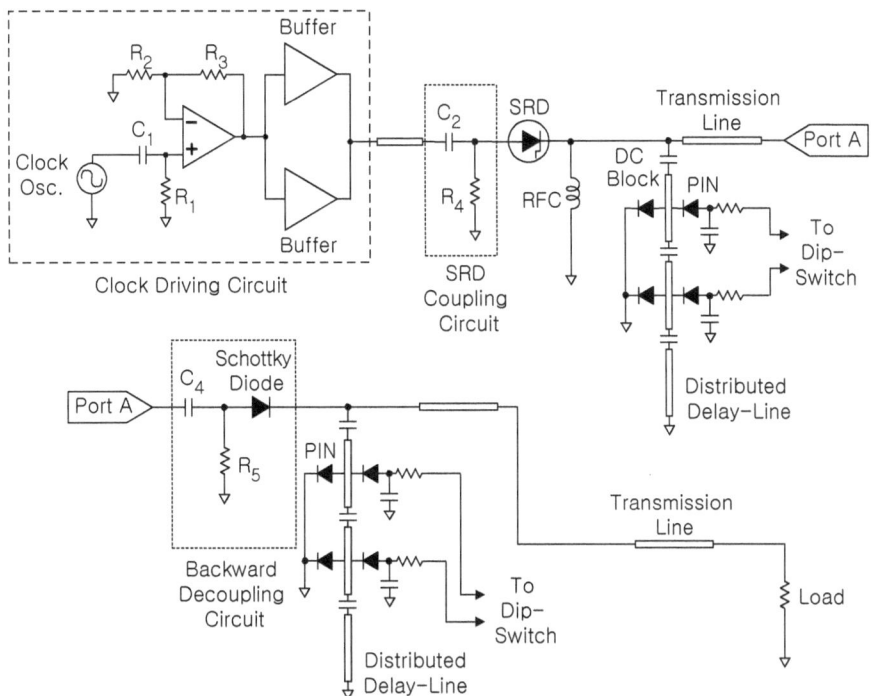

Fig. 3.5 Circuit diagram for the designed tunable monocycle pulse generator

one. It also functions as a pulse-clamping circuit. Figure 3.6a shows a simplified circuit diagram used for the design of the decoupling circuit, which includes the de-coupling circuit, a source for the incident pulse, a delay-line, an output transmission line and a load. The incident impulse is clamped to a certain DC level as shown in Fig. 3.6b and then combined with the reflected impulse from the delay-line to make a composite signal of monocycle waveform. Part of the reflected signal, represented by region A in Fig. 3.6b, propagates to the input port of the decoupling circuit in the backward direction. In order to reduce the ringing on the composite monocycle pulse, region A should be reduced. However, excessive reduction of this region may sacrifice pulse-amplitude enhancement. Compromise between the ringing level and pulse amplitude is therefore needed in the design.

Design of the backward decoupling circuit can be performed based on the well-known pulse-clamping circuit theory [26]. Assuming the input pulse is rectangular with 200-ps pulse duration and 12.5-MHz PRF, it's straightforward to draw the clamped output pulse as shown in Fig. 3.7. In Fig. 3.7, A_f and A_r represent the areas under the positive and negative voltages, respectively, and are related by the following relationship:

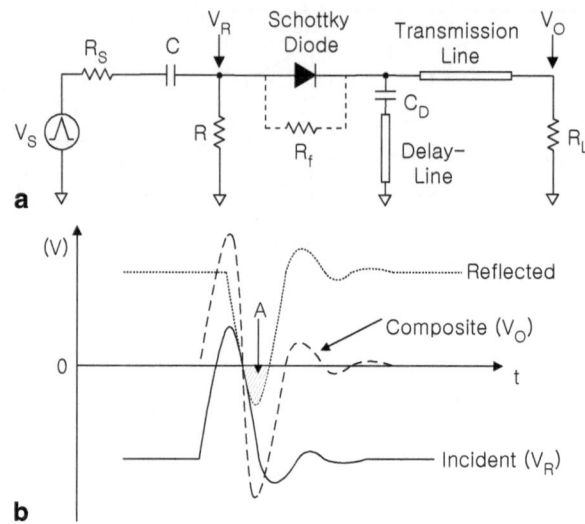

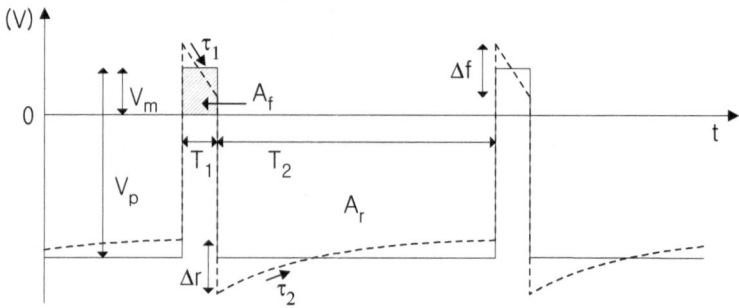

Fig. 3.7 Clamped impulse signal by the decoupling circuit with the circuit design parameters.
The *solid-* and *dashed-line* waveforms represent ideal rectangular and actual distorted pulses,
respectively

$$\frac{A_f}{A_r} = \frac{R_f}{R} \tag{3.12}$$

where R_f is the forward-diode resistance, which can be approximated by the series resistance of the Schottky diode, and R is the clamping resistance shown in
Fig. 3.6a. In our design, the pulse duration is relatively small compared to the period, resulting in a low duty cycle for the pulse generator. As implied in (3.12),
for a low duty-cycle pulse signal, it is not practical to clamp the peak of the pulse
to a small positive level close to zero. Therefore, assuming that $R_f = 10\ \Omega$ and the
clamping level V_m is just a quarter of the pulse amplitude V_p as shown in the ideal
pulse waveform in Fig. 3.7, the required $R = 10\ K\Omega$ can be calculated using (3.12).

The actual steady-state output pulse waveform of the clamping circuit has some distortion as shown in Fig. 3.7. The distortion parameter Δf has the following relationship with Δr [26]:

$$\Delta f = \frac{R_f}{R_f + R_s} \frac{R + R_s}{R} \Delta r \qquad (3.13)$$

where R_s is the source resistance. In our case, $R_s = 0$, resulting in $\Delta f = \Delta r$. Therefore, as shown in Fig. 3.7, in order to increase Δf, $\tau_2 = RC$, and hence C, should be decreased. Larger Δf produces larger amplitude for the monocycle pulse but also generates more ringing signal because of the increased portion of the backward coupling. As a compromise to achieving this contradictory objective, it is deemed reasonable to choose T_2, defined in Fig. 3.7, as RC or RC/2, from which $C = 8 \sim 16$ pF can be obtained as design values. To achieve small ringing signal, $C = 15$ pF is finally used in our design.

3.4   Fabrication and Measurement

The delay-line SRD impulse generator was designed, based on the circuit diagram of Fig. 3.1, and fabricated first in order to verify the design concept. This designed impulse generator can be used for other applications such as TDR (Time-Domain Reflectometry) applications as well. The circuit was implemented as an MIC using the microstrip structure on an RT/duroid 6010 substrate having a relative dielectric constant of 10.2 and a thickness of 0.127 cm. For the convenience of measurement, the clock driving circuit was fabricated on a separate circuit board. The clock oscillator used is the low-cost voltage controlled crystal oscillator, CSX-750VC manufactured by Citizen, which has 12-MHz PRF, 5-V CMOS logic output, and 5-ns rise and fall time. The opamp used in the clock driving circuit is THS3001 manufactured by Texas Instruments Inc., which has 420-MHz unity-gain bandwidth and 100-mA output driving capacity. The buffer is BUF634 manufactured by Burr-Brown Co., which has 180-MHz bandwidth and ±250-mA output current capacity. In the SRD coupling circuit, $R = 12~\Omega$, which is the same as the design value, and $C = 300$ pF, which is chosen as an optimal value through some experiments rather than using 200-pF design value. For the impulse shaping circuit, the SRD is SMMD-840, manufactured by Metelics Co., which has 70-ps transition time, 10-ns MCLT and 15 V of breakdown voltage. The delay-line is designed to have 100-ps round-trip time. The output pulse from the fabricated circuit is measured by an HP54750A digitizing oscilloscope with 12.4 GHz bandwidth, and the waveform is shown in Fig. 3.8. The measured impulse has 8-V peak amplitude, 160-ps FWHM (Full Width Half Maximum), and 300-ps pulse width (defined at 10% of the peak amplitude). The bias voltages used for the designed impulse and tunable monocycle pulse generators are summarized in Table 3.1.

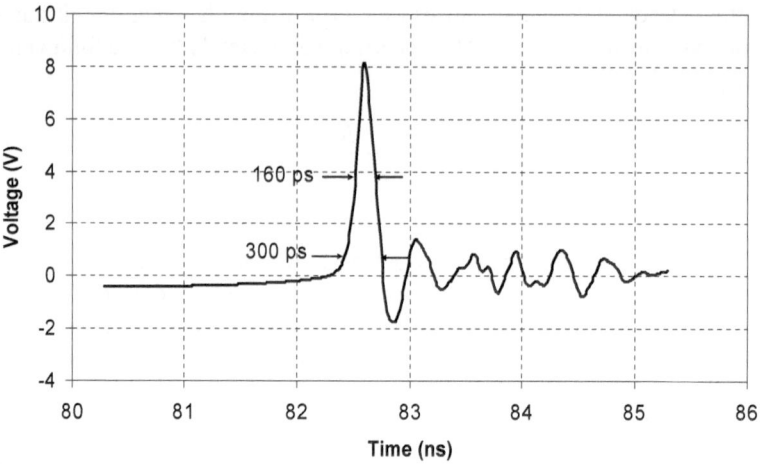

Fig. 3.8 Measured output pulse signal of the fabricated impulse generator

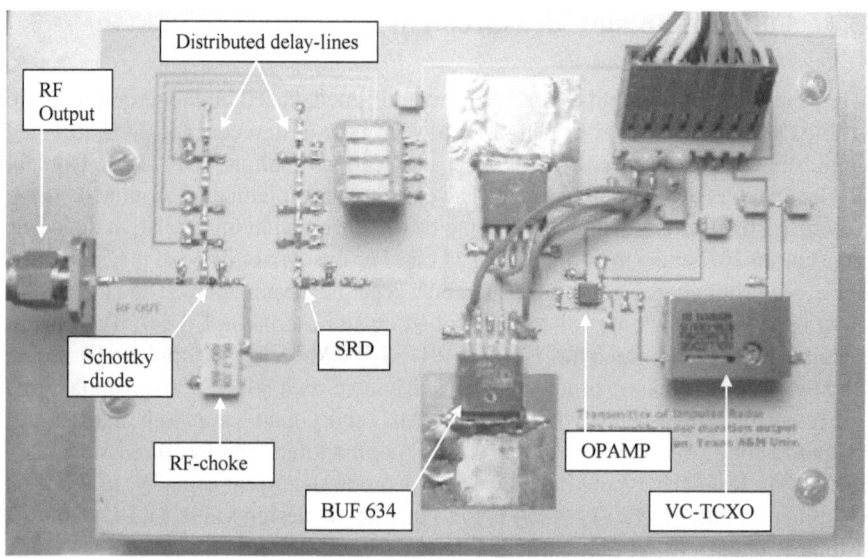

Fig. 3.9 Photograph of the fabricated transmitter of the impulse radar

The tunable monocycle pulse transmitter is also fabricated using microstrip structure on the same substrate as that used for the impulse generator. Its photograph is shown in Fig. 3.9. The clock oscillator used in this circuit is the low-cost VC-TCXO (Voltage Controlled—Temperature Controlled Crystal Oscillator) GTXO-536V manufactured by Golledge Electronics, which has 10-MHz PRF, 5-V

Table 3.1 DC bias voltages (in V) for the designed impulse and tunable monocycle pulse generators

	Clock osc.	Opamp		Buffer		PIN Diode
	V_{DD}	V_{cc^+}	V_{cc^-}	V_{cc^+}	V_{cc^-}	
Impulse generator	5	16	−5	18	−18	N/A
Monocycle generator	5.2	14	−5	18	−18	5

Table 3.2 Summary of the design parameters for the distributed delay-lines and the measured performance of the tunable monocycle pulse transmitter

DIP S/W setting	Delay-line length (inch)	Designed round-trip time (ps)	Designed pulse width (ps)	Measured pulse width@10% (ps)	Operating frequency band (GHz)	Pulse amplitude (Vpp)
001	0.220	120	410	450	0.40–3.70	5.8
010	0.400	200	570	600	0.30–2.60	8.8
100	0.700	350	870	880	0.20–1.80	9.8
000	1	500	1170	1170	0.15–1.30	9.4

HCMOS logic output, frequency (or PRF) adjustment through the voltage control and a mechanical trimmer. The reason why the VC-TCXO is used, which is related to the designed synchronous sampling receiver of the UWB system described in this book, will be explained in detail as part of the receiver design in Chap. 4.

For the SRD coupling circuit, $C = 1$ nF is chosen as an optimal value through some experiments rather than 300 pF as used in the impulse generator. This change may be considered as a result of a slight difference in the circuit structure in the pulse shaping circuitry as compared with the impulse generator. The same SRD for impulse generator is also used here. The switching p-i-n diodes used for this design are SMP1320–079 manufactured by Alpha Ind. The DC block capacitors for the decoupling of the PIN diode bias are C08 series 20-GHz DC block manufactured by Dielectric Laboratories Inc. The Schottky diode used for the backward decoupling circuit is MSS50048-E25 manufactured by Metelics. The required DC bias voltage settings for the designed transmitter is shown in Table 3.1.

The designed tunable monocycle pulse transmitter employs two identical distributed delay-lines, each divided into four sections to generate four different monocycle pulse durations. By turning on a branch of DIP switches, the corresponding two p-i-n-diode pairs located on the two distributed delay-lines are turned on simultaneously, generating an output monocycle pulse with a particular duration. The pulse duration, corresponding to the delay-line length, can be varied according to the selected turn-on position of the p-i-n-diode pairs. Table 3.2 shows the design parameters and results obtained for the tunable monocycle pulse generator. In the DIP switch setting, "1" represents connection of the corresponding branch of the switch and "0" represents disconnection. The designed pulse width, T_p, in Table 3.2

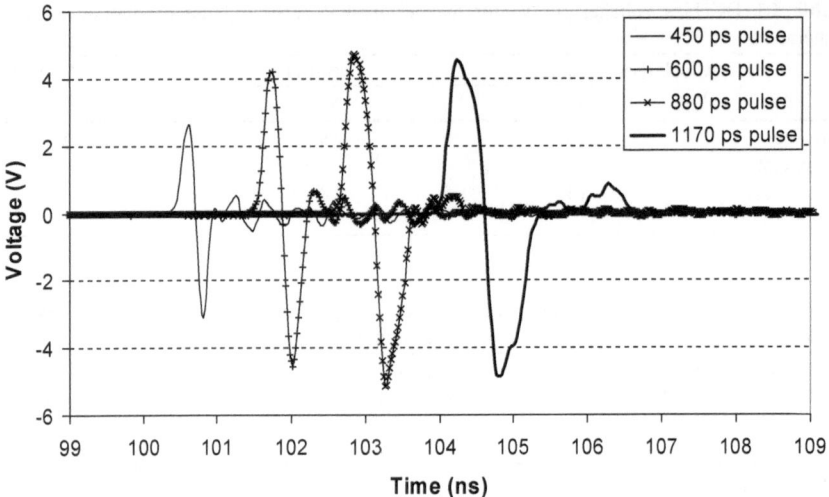

Fig. 3.10 Measured output monocycle pulses with four different pulse durations controlled by p-i-n diode switching

represents the desired pulse width of the output monocycle, which can be calculated approximately as $T_p = 2T_{rt} + t_r$, where T_{rt} is the designed round-trip time and t_r is the rise-time of the leading edge of the monocycle pulse, which is estimated as 170 ps through the impulse output in Fig. 3.8.

Figure 3.10 shows the measured output monocycle pulses with four different pulse durations. The characteristics of each generated pulse are summarized in Table 3.2 along with the designed values for comparison. The measured pulse width at 10% in Table 3.2 is the pulse width measured at about 10% level of the peak amplitude.

It is noted for reference that, according to the system analysis in Chap. 2 for the considered multi-layer pavement structure, whose parameters are listed in Tables 2.1 and 2.2, the required UWB transmitting pulse's duration is about 400 ps and the peak power should be greater than 80 mW which is corresponding to 4 V_{p-p} in voltage amplitude of the pulse. The narrowest pulse seen in Fig. 3.10 approximately satisfies this requirement.

Spectrum analysis has also been done for the generated monocycle pulse signals to determine their frequency bandwidths. Figure 3.11 shows the frequency-spectrum data for the measured 450-ps-duration monocycle pulse. The data for the frequency spectrum of the ideal 450-ps-duration monocycle pulse is also included in Fig. 3.11 for comparison. The ideal monocycle is represented as a perfect single cycle of sinusoid. In the figure, we can see that both of the main-lobe and side-lobes of the ideal monocycle pulse is higher than those of the measured actual pulse. The reason is that the ideal pulse has abrupt transitions at the beginning and ending parts of the single-cycle duration. The available bandwidth (BW) is defined in Fig. 3.11 as the instantaneous bandwidth at 10-dB level, which is known to be a

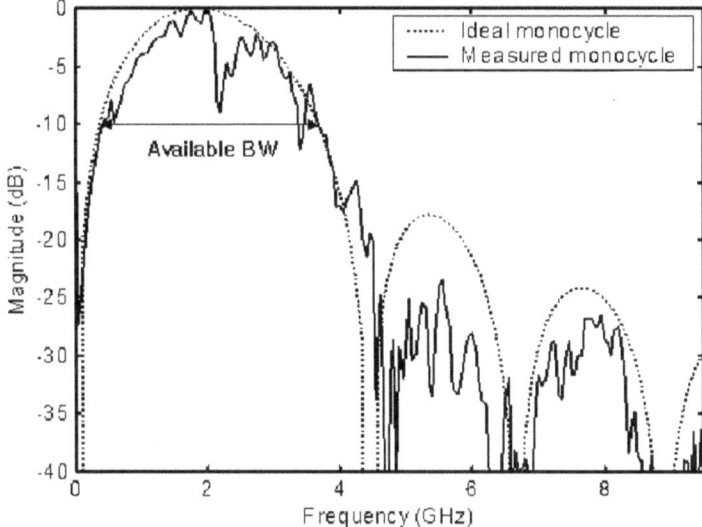

Fig. 3.11 Spectrums of the ideal and measured monocycle pulses with 450-ps duration

useful measurement to characterize the range resolution of impulse radars [1]. For instance, the 400-ps-duration monocycle pulse has about 4 GHz of available BW, which corresponds to the range resolution of 1 inch. UWB systems having more than 3-GHz available BW for the transmitting pulse signal is usually known as high-resolution systems, which can be used for high-resolution applications such as land-mine or UXO detection. Available BW data is also useful from system design aspect and needed for the design of system's components. Both the transmitting and receiving antennas need to be designed to cover the entire available BW of the transmitting pulse signal. The receiver operating BW should also be large enough to cover the available BW of the pulse to minimize the conversion loss, or maximize the conversion gain, and the distortion of the received signal and hence the down-converted signal.

Spectrum analysis results for all the four generated monocycle pulses are shown in Fig. 3.12 and the measured operating frequency bands at 10-dB level are specified in Table 3.2. The operating frequency bands of these generated pulses are from 0.15 to 3.7 GHz. Measured pulse amplitudes are in the range of 6–9 Vpp as seen in Fig. 3.10, which corresponds to 200–400 mW of pulse peak-power. The smaller amplitude obtained for the 450-ps pulse is due to the fact that the designed round-trip time of 120 ps along the delay-line is smaller than the actual transition time of 170 ps; therefore the incident step pulse could not reach the peak value before the arrival of the reflected one, resulting in reduction of the impulse amplitude. The only way to improve the amplitude for the 450-ps pulse is to use a new SRD with faster transition time, and acceptable MCLT and breakdown voltage. Such a device, however, is not commercially available at present.

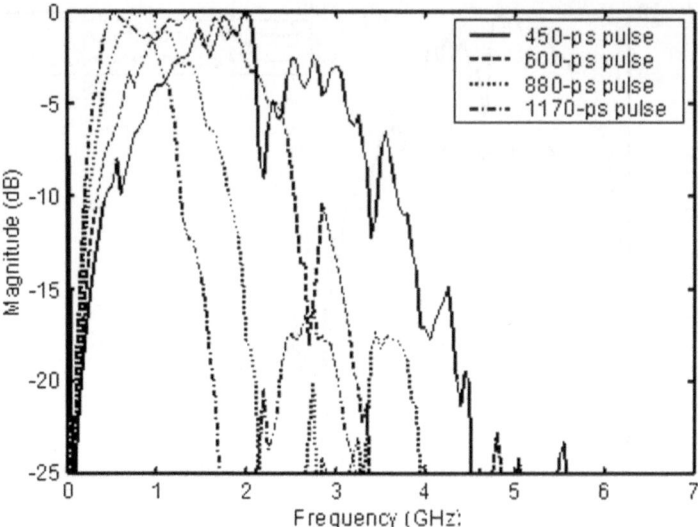

Fig. 3.12 Spectrums of all the monocycle pulses generated from the designed tunable pulse transmitter

3.5   Tunable Impulse and Monocycle Pulse Generators Implementing Switching Transistors

The tunable impulse and monocycle pulse generators described in Sects. 3.2 and 3.3 implement p-i-n diodes as the switching elements for the distributed delay-lines. Instead of p-i-n diodes, transistors such as MESFET can also be used as the switching elements. The designs and operations of these p-in-diode and MESFET pulse generators are essentially the same. However, in contrast to the PIN diodes, the MESFET is biased through its gate terminal separated from the short-circuited stub, thereby eliminating the need of a DC blocking capacitor for each MESFET/transmission-line section. The MESFET tunable pulse generator is thus simpler and produces possibly less-distorted pulses in comparison with its p-i-n-diode counterpart.

For comparison, we show in Figs. 3.13 and 3.14 the photographs and measured results of the p-i-n-diode and MESFET tunable impulse generators designed based on the approach described in Sects. 3.2 and 3.3. The employed transmission lines are microstrip lines on FR-4 glass epoxy substrate having a relative dielectric constant of 4.5 and a thickness of 0.031 in. The SRD used is MMD-0840 manufactured by Metelics Company. The p-i-n diodes used are SMP1320–079 made by Alpha Industries and the MESFETs are NE76084 manufactured by NEC. Five p-i-n-diode (MESFET)/transmission-line sections are used. The lengths of the transmission lines are chosen to produce 100-ps incremental time delay between adjacent sections. The possible longest pulse duration is around 800 ps, including about 300 ps of the SRD's transition time. Figure 3.14 shows all the measured possible pulse waveforms produced by the fabricated tunable impulse generators. The p-i-n-diode

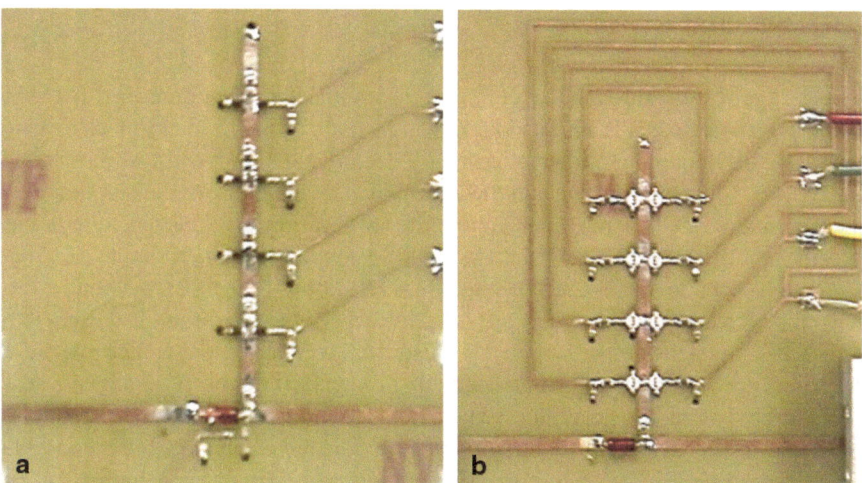

Fig. 3.13 Photograph of the p-i-n-diode **a** and MESFET **b** tunable impulse generator

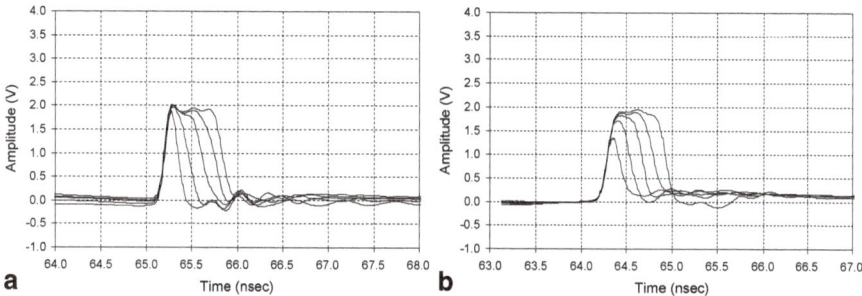

Fig. 3.14 Measured pulses of the p-i-n-diode **a** and MESFET **b** tunable impulse generator

impulse generator has measured pulse durations from 300–800 ps and amplitudes around 1.9 V. The MESFET impulse generator also exhibits measured durations from 300–800 ps with amplitudes between 1.4–1.9 V. The measured pulses of the p-i-n-diode impulse generator show reasonably constant amplitudes with no severe distortion while those of the MESFET impulse generator have little distortion. The shape of the pulses produced by the MESFET impulse generator is more symmetrical as compared to that produced by the PIN-diode pulse generator. This is due to the fact that DC blocking capacitors are not used in the MESFET impulse generator. The MESFET impulse generator also produces pulses of better symmetry and distortion than the p-i-n-diode impulse generator. Good symmetry and low distortion in pulses are important for most UWB applications.

3.6   Summary

The design of a tunable monocycle pulse transmitter has been described for low-power, short-range UWB applications. The transmitter is made up of a clock driving circuit, a SRD coupling circuit, a backward decoupling circuit and two distributed delay-lines. The clock driving and SRD coupling circuits improve the output power and transition speed. Using opamp and buffer IC's for the clock driving circuit has greatly simplified the circuit design. In the SRD coupling circuit, a simple RC filter structure is used to achieve near optimal bias condition for the SRD without external, complicated bias control circuit. A backward decoupling circuit is designed to reduce ringing in the output monocycle pulse. Tuning of the output monocycle pulse duration is achieved by alternately switching on and off the p-i-n-diode pairs spatially located along the delay lines. The employed tuning method is easy to implement and results in a compact circuit structure. The developed tunable monocycle pulse transmitter achieves varying pulse duration from 0.4 to 1.2 ns, corresponding approximately to the operating frequency range of 0.15–3.7 GHz, and 200–400 mW of peak power. These results show that the designed monocycle transmitter with advanced tuning capability can be used for most short-range UWB applications, even for high-resolution radar applications such as UXO and land-mine detection. The impulse generator, developed along with the tunable monocycle pulse transmitter, exhibits a performance of 160-ps FWHM and 8-V peak amplitude and can also be used for UWB systems. Comparison between the tunable impulse generators employing p-i-n diodes and MESFETs as the switching elements shows the MESFET impulse generator produces relatively less distorted and more symmetric pulses than its p-i-n-diode counterpart. The MESFET tunable impulse generator also does not require DC blocking capacitors and is thus simpler.

References

1. Yarovoy, A., Ligthart, L.: Full-polarimetric video impulse radar for landmine detection: Experimental verification of main design ideas. In: Proceedings of the 2nd International Workshop on Advanced Ground Penetrating Radar, pp. 148–155 (2003)
2. Miller, E.K.: Time-Domain Measurements in Electromagnetics. Van Nostrand Reinhold, New-York (1986)
3. Nicolson, A.M.: Subnanosecond risetime pulse generators. IEEE Trans. Instrum. Meas. **25**(2), 104–107 (June 1976)
4. Sayadian, H.A., Li, M.G., Lee, C.H.: Generation of high-power broad-band microwave pulses by picosecond optoelectronic technique. IEEE Trans. Microw. Theory Techn. 37(1), 43–50 (January 1989)
5. Forcia, R.J., Schamiloglu, E., Fleddermann, C.B.: Simple techniques for the generation of high peak power pulses with nanosecond and subnanosecond rise times. Rev. Sci. Instrum. 67(7), 2626–2629 (July 1996)
6. Fontana, R.J., Richley, E.A., Beard, L.C., Barney, J.: A programmable ultra wideband signal generator for electromagnetic susceptibility testing. In: 2003 IEEE Conference on Ultra Wideband Systems and Technologies, pp. 21–25 (2003)

7. Andrews, J.R.: Picosecond pulse generators for UWB radars. Picosecond Pulse Labs, Boulder, CO, Application Note AN-9, May 2000

8. Daneshvar, K., Howard, L.: High current nanosecond pulse generator. In: Proceedings IEEE Southeastcon '89, pp. 572–576 (1989)

9. Uhmeyer, U.A., Libby, J.C.: A fast variable transition time pulse generating circuit. In: Proceedings of the 9th IEEE Instrumentation and Measurement Technology Conference, pp. 152–157 (1992)

10. Han, J., Nguyen, C.: Ultra-wideband electronically tunable pulse generators. IEEE Microw. Wirel. Compon. Lett. 14(30), 112–114 (March 2004)

11. Lee, J.S., Nguyen, C.: A uniplanar picosecond pulse generator using step-recovery diode. Electron. Lett. **37**(8), 504–506 (April 2001)

12. Lee, J.S., Nguyen, C.: New uniplanar subnanosecond monocycle pulse generator and transformer for time-domain microwave applications. IEEE Trans. Microw. Theory Tech. **49**(6), 1126–1129 (June 2001)

13. Lee, J.S., Nguyen, C.: Novel low-cost ultra-wideband, ultra-short-pulse transmitter with MESFET impulse-shaping circuitry for reduced distortion and improved pulse repetition rate. IEEE Microw. Wirel. Compon. Lett. **11**(5), 208–210 (May 2001)

14. Han, J., Nguyen, C.: A new ultra-wideband, ultra-short monocycle pulse generator with reduced ringing. IEEE Microw. Wirel. Compon. Lett. **12**(6), 206–208 (June 2002)

15. Lee, J.S., Nguyen, C., Scullion, T.: New uniplanar subnanosecond monocycle pulse generator and transformer for time-domain microwave applications. IEEE Trans. Microw. Theory Techn. **49**(6), 1126–1129 (June 2001)

16. Han, J., Nguyen, C.: Ultra-wideband electronically tunable pulse generators. IEEE Microw. Wireless Compon. Lett. **14**(3), 112–114 (March 2004)

17. Han, J.W., Nguyen, C.: On the Development of a Compact Sub-Nanosecond Tunable Monocycle Pulse Transmitter for UWB Applications. IEEE Trans. Microw. Theory Tech. **MTT-54**(1), 285–293 (January 2006)

18. Miao, M., Nguyen, C.: On the Development of an Integrated CMOS-Based UWB Tunable–Pulse Transmit Module. IEEE Trans. Microw. Theory Tech. **MTT-54**(10), 3681–3687 (October 2006)

19. Lesha, M.J., Paoloni, F.J.: Generation of balanced subnanosecond pulses using step-recovery diodes. Electron. Lett. **31**(7), 510–511 (March 1995)

20. Pulse and waveform generation with step recovery diodes, Agilent Technologies Inc., Palo Alto, CA, Application Note 918 (1968)

21. Cormack, G.D., Sabharwal, A.P.: Picosecond pulse generator using delay lines. IEEE Trans. Instrum. Meas. **42**(5), 947–948 (October 1993)

22. Hamilton, S., Hall, R.: Shunt-mode harmonic generation using step recovery diodes. Microw. J. 69–78 (April 1967)

23. Moll, J.L., Hamilton, S.: Physical modeling of the step recovery diode for pulse and harmonic generation circuits. Proc. IEEE. **57**(7), 1250–1259 (July 1969)

24. Goldman, S.: Computer aids design of impulse multipliers. Microw. RF. 101–128 (October 1983)

25. Zhang, J., Räisänen, V.: Computer-aided design of step recovery diode frequency multipliers. IEEE Trans. Microw. Theory Tech. **44**(12), 2612–2616 (December 1996)

26. Millman, J., Taub, H.: Pulse, Digital, and Switching Waveforms. McGraw-Hill, New-York (1965)

Chapter 4
UWB Receiver Design

4.1 Introduction

The receiver of UWB systems is used to down-convert UWB signals received from the receiving antenna to baseband signals. Perhaps the most challenging problem in the design of the UWB receiver is down-converting such a wideband input signal while still retaining the down-converted signal waveform in the same form as the input signal. Achieving this feature would mean the UWB receiver has high signal fidelity in the signal conversion over an ultra-wide bandwidth which is probably the utmost important characteristic of the UWB receiver.

Unlike continuous-wave (CW) based systems, the synchronous sampling (or equivalent time sampling) receiver structure is commonly used for UWB systems [1], [2], [3–6]. Other receiver architectures for UWB systems such as the channelized ADC (Analog-to-Digital Conversion) receiver are also employed [7]. Among them, however, the synchronous sampling method is much simpler and provides more compact systems. Commercially available wideband sampling oscilloscopes can be used as UWB receivers; yet they are bulky, expensive, and not suitable for practical UWB systems, particularly those require mobility, portability, low cost and/or small operating platforms.

The synchronous sampling method has been widely used in electro-optic sampling techniques to down-convert RF signal or to reproduce fast transient signal on a large time scale [4, 5]. Figure 4.1 illustrates the basic principle of the synchronous sampling method. The synchronous sampling in time domain, in principle, is basically the same as the harmonic mixing of two signals of different frequencies in frequency domain. In Fig. 4.1, V_R represents the waveform of the input RF signal and V_O is the local oscillator (LO) strobe pulse signal that triggers the sampling. V_D is the output down-converted signal through the (assumed ideal) sample-and-hold operation of the receiver circuit. As can be inferred from Fig. 4.1, down-converting a pulse signal is electrically equivalent to expanding or stretching the pulse itself or, in other words, reproducing a pulse of narrow (e.g., sub-nanosecond) duration into a much wider (e.g., sub-millisecond) pulse. To achieve down-conversion of the RF input V_R, the frequency of V_R and PRF of V_O should have a small difference between them. For UWB systems, V_R is actually a pulse

C. Nguyen, J. Han, *Time-Domain Ultra-Wideband Radar, Sensor and Components*, 47
SpringerBriefs in Electrical and Computer Engineering,
DOI 10.1007/978-1-4614-9578-9_4, © Springer International Publishing Switzerland 2014

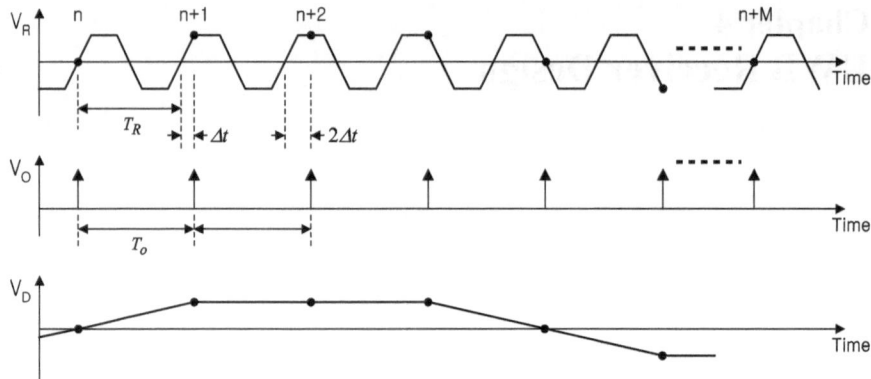

Fig. 4.1 Illustration of the principle of synchronous sampling down-conversion

signal having a PRF, f_R, and an extremely large duty cycle. For down-converting the RF pulse, the PRF of the LO strobe pulse, f_o, should have a small deviation from f_R; therefore, the synchronous sampling receiver of UWB systems should have two stable reference oscillators with a small frequency deviation between them. One oscillator is used in the transmitter and determines the PRF of the transmitting pulse, which is also the PRF of the RF input pulse to the receiver. The other oscillator is used for the LO strobe pulse generator of the receiver to determine the PRF of the strobe pulse.

There is another method for realizing the synchronous sampling receiver other than using two reference oscillators with different frequencies. Two identical oscillators with the same frequencies can be used if proper timing of the sampling can be provided by a timing control circuit. In Fig. 4.1, the dots on the V_R waveform represent the sampled points numbered from n to n + M. Notice that each pair of two successive sample points have the same time delay of Δt from the RF input signal period T_s. Therefore, even in the situation that the PRF of the strobe pulse is the same as that of the RF input pulse, if a constant time delay Δt occurs for the next period of the signal, correct timing of the sampling can be achieved. The successive time delay generation can be implemented by using the digital delay generator, which is a digital IC commonly used in the timing control of digital circuits. The synchronous sampling using the delay generator can be implemented at low cost, whereas the receiver structure is more complicated than that consisting of using two oscillators having slightly different frequencies. Other disadvantages of using a delay generator are that minimum delay time realizable is limited to 10 ps with currently available commercial products and the number of sample points of a down-converted signal during a time interval of a single clock period is also limited by the internal code length (usually 8-bit) of the delay generator.

To implement the UWB receiver with more compact circuit structure and avoid the limitations of the delay generator, the synchronous sampling method using two oscillators with a small frequency difference is chosen. Figure 4.2 shows the UWB receiver based on the synchronous receiver architecture, along with the UWB

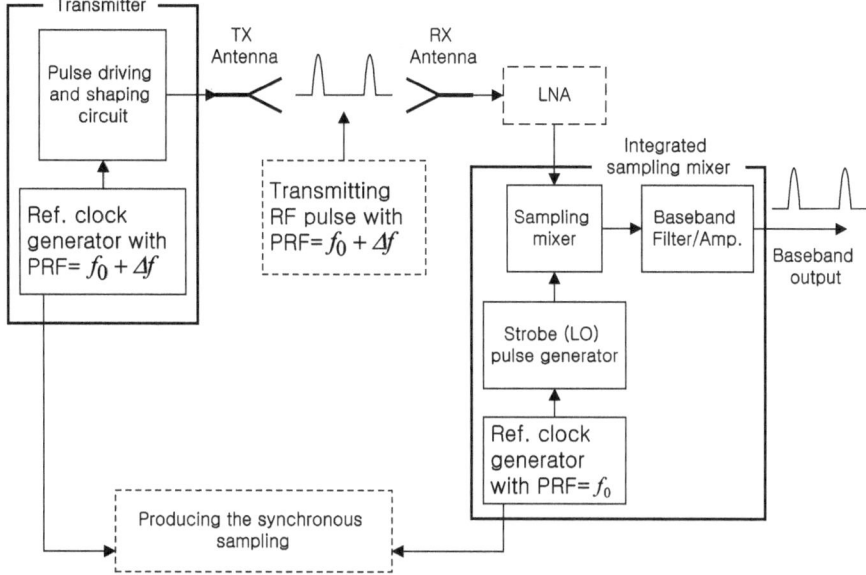

Fig. 4.2 Block diagram of the UWB receiver along with the UWB transmitter for the UWB system. The TX and RX antennas do not point to each other in actual operation

transmitter, for the UWB system presented in this book. The synchronous sampling is implemented by using two reference clock oscillators: one for the receiver and another for the transmitter, with a small frequency deviation of Δf between them. As indicated in Fig. 4.2, the UWB receiver is composed of a low-noise amplifier (LNA) and an integrated sampling mixer.

It is noted that the LNA as seen in Fig. 4.2 is only required if the received signal from the receiving antenna is too weak to be detectable or triggered in the baseband circuit. In some applications, the UWB transmitter does not need to supply maximum peak-power, so much smaller peak-power can be used in a low-power mode in the transmitter to save the DC power consumption in the system. When the UWB transmitter is used in a low-power mode, the LNA may be needed to amplify a weak received signal. On the other hand, using the UWB transmitter in the full-power mode may lead to the received signal being so large that there is no need to use the LNA to trigger or detect the received signal in the baseband circuit. Using the LNA in the full-power mode of the UWB transmitter may create problems such as potential saturation of the sampling mixer through the excessively large output of the LNA, causing distortion in the down-converted signal. This problem is in fact found in the experiments of the designed UWB system described in this book. Since the designed UWB system is intended for operation in the full-power mode for our applications such as assessment of pavements, the LNA is not included in the system. An UWB LNA suitable for the UWB receiver, however, will be described in the last section of this chapter as a reference for the LNA design for UWB systems. Particularly, the time-domain performance of this amplifier will be discussed in details to show its high signal fidelity in reproducing faithfully the waveforms of input UWB pulse signals which is vital for the performance of UWB systems.

Without the LNA, the UWB receiver of the UWB system as seen in Fig. 4.2 is only the integrated sampling mixer. All of the required components for the down-conversion of the RF input pulse are the sampling mixer (or sampler), strobe pulse generator, reference clock oscillator, and baseband filter and amplifier, which compose the integrated sampling mixer as shown in Fig. 4.2. The integrated sampling mixer is so named since all the required components are integrated in a single printed circuit board (PCB) as a hybrid microwave integrated circuit (MIC).

The sampling mixer, which is the key component of the integrated sampling mixer (and hence the UWB receiver), has been widely used for wideband frequency down-conversion in many applications. One application is for microwave instrumentation such as network analyzers, frequency counters, and digitized sampling oscilloscopes [8–22]. In this application, the sampling mixer is used for synchronous sampling (or sub-sampling) of a fast transient signal to recover it on a large time scale or for down-conversion of a CW signal. Solid-state millimeter-wave and electro-optic samplers, which can sub-sample picosecond transient signals, have also been developed for instruments operating in the millimeter-wave band [11–21]. Another important application of the sampling mixer is for UWB systems such as that used in this book, which employ sub-nanosecond pulse as the transmitting signal [1, 3, 23–25]. In UWB systems, such as subsurface penetrating radar, sub-sampling of the received signal by a sampling mixer is needed to extract detected information [26]. Sampling mixers for UWB applications particularly require low conversion loss and high dynamic range because of its direct conversion operation in the UWB receiver. Some analyses have also been reported for the sampler [27, 28].

In this chapter, we present the design of an UWB receiver based on synchronous sampling that consists of a sampling mixer integrated with a local strobe-pulse generator and a baseband filter and amplifier. A simple equation allowing accurate prediction of the sampling mixer's intrinsic conversion loss and RF operating bandwidth is derived. Development of the strobe pulse generators operating at such low clock frequency as 10 MHz is also presented along with the design information for circuit components. The developed sampling mixer is called coupled-slotline-hybrid (CSH) sampler stemming from its use of a CSH. Double-sided planar structure is used for low-cost fabrication and ease of integration with the strobe-pulse generator and the baseband circuit using hybrid MIC technique. The CSH sampler employs a modified version of the coupled-slotline magic-T [29] for the coupling and termination of the RF and LO signals to the sampling-bridge circuit. The modified CSH allows the sampling bridge to be included and suppression of the ringing of the strobe pulse signal to be achieved. The developed CSH sampler achieves a conversion loss of 4.5–7.5 dB (without amplifier) and conversion gain from 6.5–9.5 dB (with amplifier) over a 3-dB bandwidth of 5.5 GHz. The sampler has a dynamic range greater than 50 dB and a sensitivity of -47 dBm.

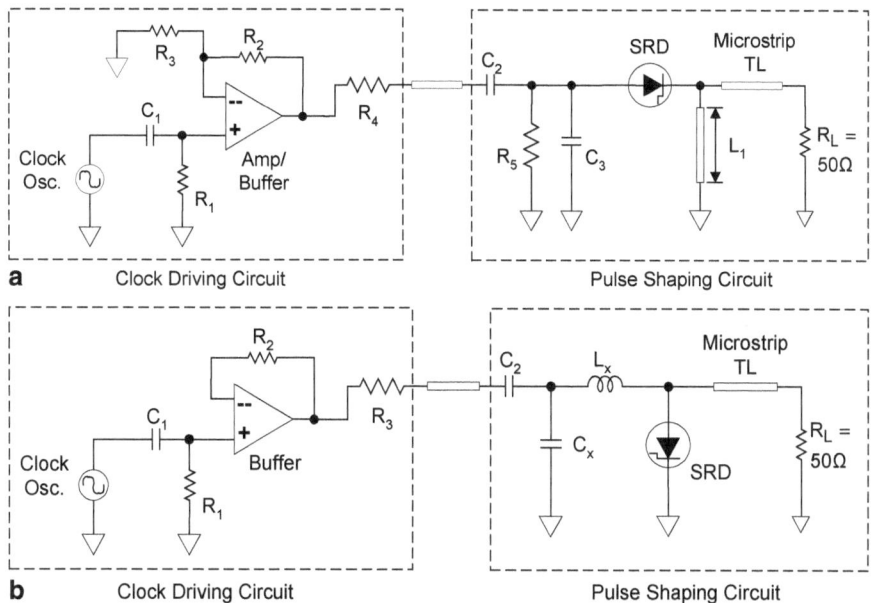

a   Clock Driving Circuit   Pulse Shaping Circuit

b   Clock Driving Circuit   Pulse Shaping Circuit

Fig. 4.3 Circuit diagrams for **a** the delay-line, and **b** shunt-mode pulse generators

4.2   Design of Strobe Pulse Generator

SRD (Step Recovery Diode) impulse generators have been used for generation of sub-nanosecond strobe pulses for samplers operating up to more than 10 GHz [13–15]. Design methods for SRD impulse generators are well developed and used for various multiplier and sampler designs [30–33]. These design methods, however, are applicable for high (input) driving-source or clock frequencies. They do not provide optimized and good performance when the driving-source frequency is low (e.g., 10 MHz). Low driving frequency is needed for various applications. In UWB systems requiring sub-sampling in the receiver, the transmitted pulse's PRF is determined by the strobe pulse's PRF in the receiver, which may be as low as 10 MHz in order to have sufficient observation time and detection range.

Two different types of SRD impulse generators were designed with good performance at low driving frequencies based on well-known circuit configurations [30, 34]. One is the delay-line type and the other is the shunt-mode type. These two types were considered because the former can generate short pulse duration and the latter can produce high pulse power. Designed circuit diagrams, including the clock driving circuits, for these two different types are shown in Fig. 4.3. These pulse circuits were optimized to obtain high pulse power for low PRF with no external DC bias required for normal operation.

Since the shunt-mode pulse generator has better performance than its delay-line counterpart for sampler application and part of their designs is common, only the

Table 4.1 Specification summary of the clock oscillator and SRD used in the pulse generator design

Clock oscillator	Rise-time	Max. 5 ns
	PRF	12 MHz
	Output voltage level	5 V
SRD	Minority carrier life-time	10 ns
	Turn-off transition time	70 ps

design of the shunt-mode pulse generator is described here. The measured performance of both types, however, will be presented.

In the shunt-mode pulse generator, as shown in Fig. 4.3b, a clock oscillator is used as the driving source for the pulse-shaping circuit. To obtain good output power efficiency and high pulse amplitude, the frequency of the driving source should be less than the minority carrier life-time of the SRD. In the case of a 10-MHz driving source, an SRD having at least 100-ns life-time must therefore be used. However, in reality, this kind of long life-time SRD can not have fast transition time. To alleviate this problem and satisfy both the conditions of life-time and transition time, a fast clock signal source, whose rise-time is comparable to the SRD's life-time, is used. Summarized specifications of the selected clock oscillator and the SRD for our design are shown in Table 4.1.

In Fig. 4.3b, C_1 and R_1 are required for AC coupling of the source to the buffer. The buffer is used to provide a good impedance matching between the oscillator and pulse shaping circuit and to supply a sufficient current to the load. The buffer used is an opamp, which has a wide bandwidth larger than 150 MHz to maintain the fast rise-time of the input clock signal. It has a maximum output current level of 100 mA. The resistor R_3 is required to stabilize the circuit operation when the load of the buffer is in low impedance state. The pulse-shaping circuit is basically a shunt-mode SRD impulse generator [30]. C_2, SRD, and R_L make up the pulse clamping circuit, which functions as a self-biasing network to provide proper bias required for the shunt-mode pulse generator. The driving inductance L_x was determined approximately for initial design value using the equations in [30], and 1 nH was chosen to include some design margin. The capacitor C_x is needed to form a RF short-circuit at the time of the diode turn-off so that L_x, C_d (depletion capacitance of SRD), and R_L form a parallel-resonant ringing circuit required for generation of the impulse. Therefore, C_x should be large enough compared to C_d such that it has a low impedance value over the pulse frequency band. However, the combination of C_x and the equivalent resistance of R_3 and R_L results in a low-pass filter for the input driving signal so that C_x may not exceed some certain value to prevent the input clock signal from slewing. Therefore, $C_x < t_r/(2.2 R_e)$, where t_r is the rise-time of the clock signal and R_e is a parallel combination of R_3 and R_L. In the case of $t_r = 2.5$ ns, C_x is found to be less than 113 pF. Experiment with the fabricated circuit shows that C_x from 50 to 100 pF can be used.

Figure 4.4 shows the measured output impulse waveform from the fabricated circuits for the delay-line and shunt-mode types. These waveforms were measured by a digitizing oscilloscope, whose bandwidth is 12 GHz. The full widths at half maximum (FWHMs) of the generated pulses are about 150 ps. The pulse waveform

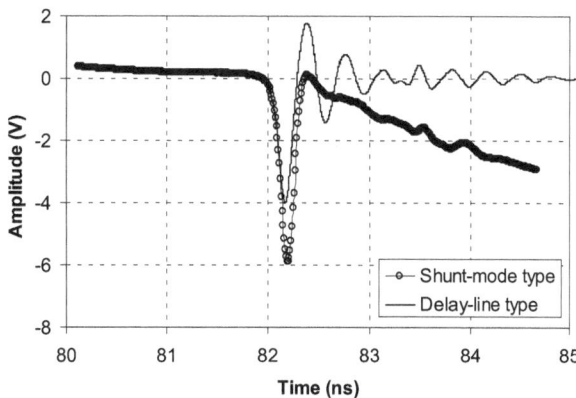

Fig. 4.4 Measured output pulse waveforms of the delay-line and shunt-mode pulse generators

of the delay-line pulse generator has large ringing even though the pulse duration is shorter than that of the shunt-mode pulse generator. Large ringing can cause some distortion on the down-converted output of the sampler. Its amplitude is also smaller than that of the shunt-mode type, while the slew rates of both pulses are the same. For pulses of the same slew rate, the actual sampling aperture times, which determine the bandwidth of the sampler, are almost the same because of the clamping effect of the sampling-bridge circuit. The clamping circuit structure of the sampling bridge and the relationship between the sampling aperture time and bandwidth of the sampler will be shown in Sect. 4.3. Furthermore, for the same slew-rate pulses, a higher amplitude pulse improves the conversion loss and increases the 1-dB compression point of the sampler. Therefore, the shunt-mode type pulse generator is a better choice for sampler in all aspects.

4.3 Coupled-Slotline-Hybrid Sampler

4.3.1 Design of the Coupled-Slotline-Hybrid Sampler

A sampler down-converts an RF signal into a baseband signal according to the principle of sampling down-conversion, which is a well–known and efficient technique of wideband down-conversion. A sampler is basically a sample-and-hold circuit in which the voltage level of an RF input signal is sampled using a sampling strobe pulse, and the sampled voltage level is maintained in the holding capacitor. Unlike common microwave mixers, a sampler works with a strobe pulse local oscillator (LO) signal rather than a continuous wave (CW) LO signal.

A sampler should have the required bandwidth and low conversion loss in the RF-to-baseband conversion. The bandwidth of the sampler should accommodate the bandwidth of the RF input signal to avoid signal distortion due to insufficient bandwidth. A major factor in determining the bandwidth of a sampler is the duration

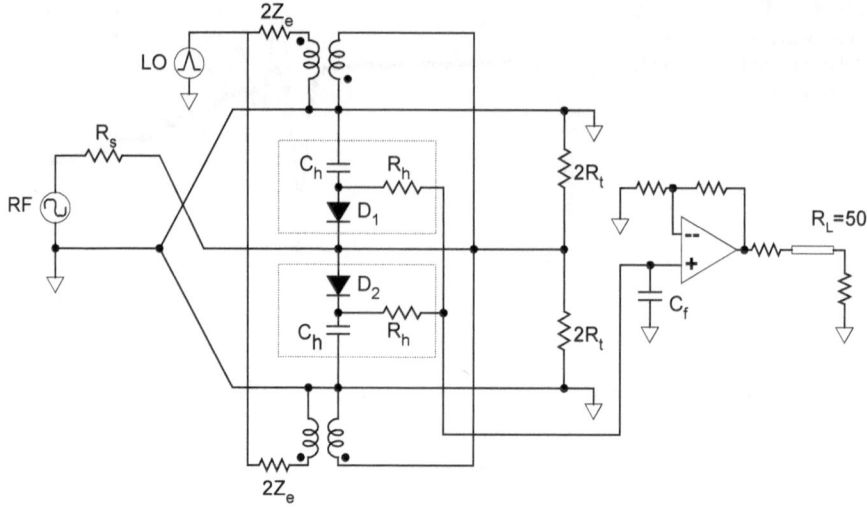

Fig. 4.5 Equivalent circuit diagram of the CSH sampler

of the strobe pulse used in the sampler. For instance, the strobe pulse generator needs to generate pulses with duration in the order of 100 ps to obtain several GHz bandwidth for a sampler. Large dynamic range for a sampler is also important for UWB systems.

The designed coupled-slotline-hybrid (CSH) sampling circuit is a two-diode-bridge configuration. Its equivalent circuit diagram, including the baseband buffer/amplifier circuit, is shown in Fig. 4.5. The equivalent circuit follows the split-ground configuration proposed in [8] for reducing the parasitic inductance of the signal transmission line.

Each part of the sampling bridge, consisting of a Schottky sampling diode (D_1 or D_2), a holding capacitor (C_h), and a holding resistor (R_h), forms a pulse-clamping circuit. The clamping circuit acts as a self-biasing network so that no external bias network is required. In the sampling circuit, C_h also works as a coupling capacitor for the LO pulse. Its capacitance should thus be sufficiently large as compared to the junction capacitance of the sampling diode in order to increase the forward bias current and decrease the diode's junction resistance, resulting in a decrease of the charging time constant of the RF signal and eventually improving the conversion loss of the sampling mixer. In another point of view, when the sampling diode is turned-off, a large holding capacitance would reduce loss of the sampled RF voltage, effectively improving the conversion loss. On the contrary, small capacitance of C_h is desirable to facilitate good RF input matching over a wide frequency range. With the junction capacitance of the beam-lead Schottky diodes used in our design of about 0.1 pF, proper value of C_h was estimated from 0.2 to 0.5 pF. Through circuit simulations of the designed sampler using Agilent ADS [35], a value of 0.2 pF was finally chosen for C_h to accomplish a wide bandwidth for the sampler.

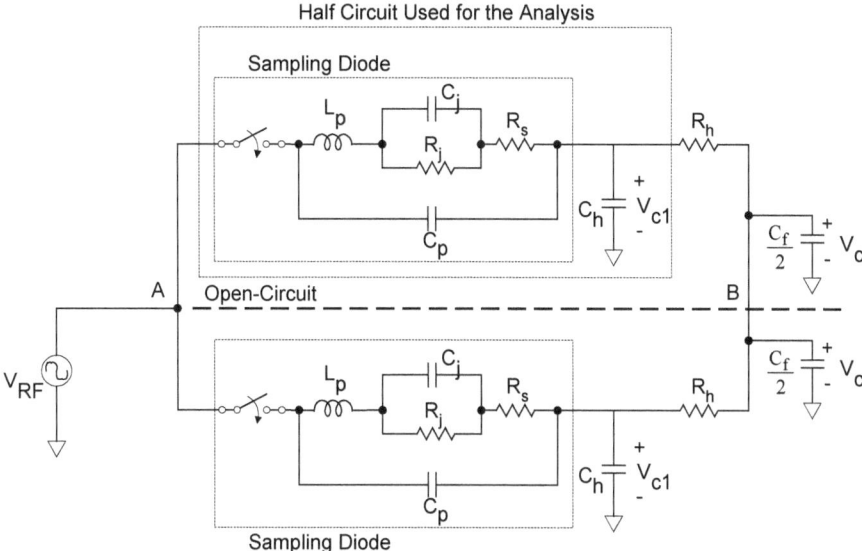

Fig. 4.6 Equivalent-circuit model of the two-diode-bridge sampler, including the baseband filter, with respect to the RF signal sampling (i.e., during the time the RF signal is sampled)

The two-diode-bridge sampling circuit is analyzed quantitatively using an equivalent circuit model, shown in Fig. 4.6, to obtain analytical result for the intrinsic conversion loss and, hence, the frequency response of the sampler. As indicated in Fig. 4.6, the dashed line connected nodes A and B represents an open-circuit considered in the analysis. Assuming identical sampling diodes and holding capacitors, the sampled voltages V_{c1} and V_{c2} are equal. The resultant sampled voltage V_c through the baseband filter is equal to V_{c1} and V_{c2} because of the resistive divider form of the voltage combining structure. Therefore, only a half of the overall equivalent circuit model (without the baseband filter structure) may be considered for the analysis. The Schottky sampling diode is represented in Fig. 4.6 as a combination of a switch and the diode's turn-on equivalent circuit model. L_p and C_p represent the diode's package parasitics. The switch can be modeled as an RC low-pass filter. The cut-off frequency of the switch is determined by the effective sampling time represented as

$$T_s = \sqrt{t_r^2 + t_a^2} \tag{4.1}$$

where t_r denotes the rise-time of the RF signal charging through the diode and t_a represents the sampling aperture time [17, 20]. In good sampling diodes, the junction capacitance is normally so small that t_r can be ignored. Using this model of the switch and a half of the equivalent circuit model in Fig. 4.6 without the baseband filter, the following equation can be derived:

$$\left|\frac{V_{c1}}{V_{RF}}\right| = \frac{1}{\sqrt{1+\left(\frac{\omega T_s}{2.2}\right)^2}} \sqrt{\frac{1+\left(\omega R_j C_j\right)^2}{\left(1-\omega^2 R_s C_h R_j C_j\right)^2 + \left\{\omega\left[R_s C_h + R_j\left(C_h + C_j\right)\right]\right\}^2}}$$

$$(4.2)$$

L_p and C_p are neglected in the derivation of (4.2) since typical good beam-lead Schottky diodes have small L_p and C_p. Eq. (4.2) calculates the intrinsic (sampling) conversion loss of the sampler due to two sampling diodes and holding capacitors. It represents a relative or normalized conversion loss with respect to the minimum conversion loss occurred at the lowest RF frequency. Here, the minimum conversion is defined as due mainly to the charged RF signal leakage during the signal charging and discharging phases through a low parasitic impedance formed by the inductive and capacitive parasitic of the holding resistor R_h and is a critical performance parameter of the sampler. Its measured value for the designed sampler is given in Sect. 4.3.2. The usefulness of (4.2) lies in the fact that it can predict accurately the conversion-loss behavior and RF operating bandwidth of the sampler, as will be seen in Sect. 4.3.2.

The sampling aperture time is dependent on the strobe pulse width and the reverse bias applied to the sampling diodes. In the designed sampling-bridge circuit, a self-bias occurs due to the clamping effect, effectively producing reverse bias to the sampling diodes. From the simulation of the circuit shown in Fig. 4.6 with an actual diode model, the sampling aperture time was estimated as 60–70 ps. Using the diode parameter values provided by the manufacturer (Rj=16 Ω, Cj=0.1 pF, Rs=11 Ω) and the estimated aperture time of 60 ps, the intrinsic conversion loss of the sampler for C_h=0.2 pF was calculated using (4.2). This conversion loss will be shown together with the measurement result in Fig. 4.11 in Sect. 4.3.2. From this analysis, the estimated RF bandwidth can be found approximately as 5.5 GHz, which is well coincided with the measurement result.

In Fig. 4.5, when the diode is turned-off, C_h, R_h and C_f compose a discharging path for the charged RF voltage in C_h, where R_h and C_f work as a low-pass filter for the baseband signal. This baseband filter is necessary to combine sampled voltages through sampling diodes and to reduce the output noise power. The required value for the holding resistor, R_h, was initially calculated as 80 KΩ by considering the strobe pulse PRF and the required time constant. However, the final value of R_h was selected as 30 KΩ to reduce the effect of its parasitic impedance on the conversion loss and to avoid an increase of the noise in the baseband signal, which is likely to occur with a large resistance. C_f was chosen as 7 pF in order to have a cutoff frequency of 1.5 MHz for the low-pass filter. R_t is a terminating resistance for both the RF and strobe pulse signals. An opamp is used as the baseband amplifier.

Figure 4.7 shows a layout of the CSH sampler, including only such major components as the CSH, sampling-bridge circuit, and connection to the baseband circuit. The basic configuration and design of the CSH follows the well-known coupled-slotline magic-T [29]. The LO pulse coupling is achieved through an underside microstrip line (rather than an air bridge) because it facilitates resistive termination required for the strobe pulse generation, is easily fabricated by a PCB manufacturing

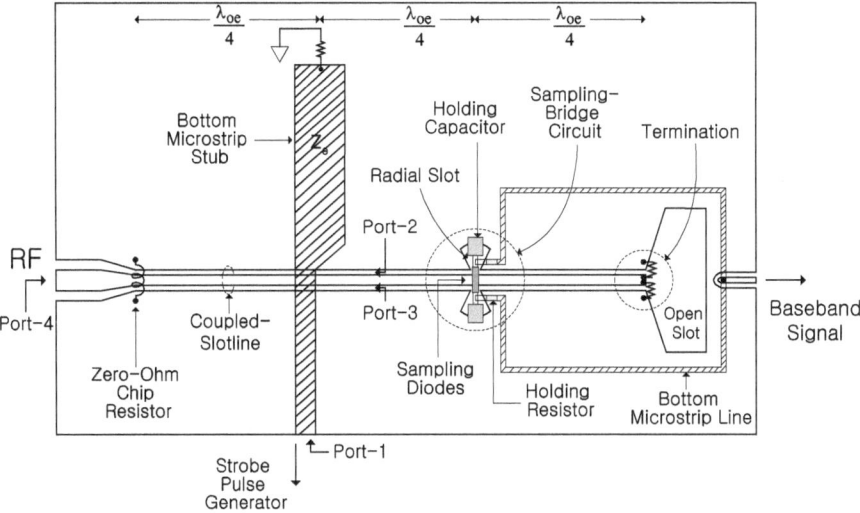

Fig. 4.7 Layout of the designed sampling circuit with main RF components. The strobe pulse generator and baseband circuits are not shown. The indicated port numbers are used for simulation purposes as shown in Fig. 4.8

process, and provides more rigid, reproducible structure for the hybrid MIC. The extended quarter-wavelength coupled-slotlines between the sampling bridge and terminations, and the open slot at the terminations help reduce the ringing in the strobe pulse. The two underside microstrip lines connecting to the sampling bridge facilitate integration with the baseband circuitry.

The even-mode (Z_{oe}) and odd-mode (Z_{oo}) characteristic impedances of the coupled-slotlines used in the CSH can be obtained, assuming no termination at the open end of the bottom microstrip stub, as

$$Z_{oe} = Z_o$$
$$Z_{oo} = 2R_t = 2Z_o$$

(4.3)

where Z_o is the characteristic impedance of the feeding CPW and microstrip line at the RF and LO ports, respectively. R_t is a half of the terminating resistor shown in Fig. 4.5. In Fig. 4.7, λ_{oe} is the even-mode wave-length at the center frequency of the monocycle pulse appeared at port 2 or 3. This center frequency is 3.3 GHz corresponding to the monocycle pulse duration of 300 ps. It should be noted here that the monocycle pulse is formed by combining two opposite-polarity impulses: half of the LO impulse signal reflected from the zero-ohm chip resistor and that propagating toward port 2 or 3. Therefore, $\lambda_{oe}/2$ is corresponding to the distance along which the strobe impulse propagates on the coupled-slotline during its duration.

The CSH was implemented using $Z_o = 50\ \Omega$, $Z_{oe} = 70\ \Omega$ and $Z_{oo} = 100\ \Omega$. 70 Ω was used for Z_{oe}, instead of 50 Ω according to (4.3), in order to have reasonable transmission-line dimensions and to accommodate the sampling bridge, using RT/ duroid 6010 substrate having a relative dielectric constant of 10.2 and a thickness of 0.127 cm. The characteristic impedance, Z_e, of the bottom microstrip stub was determined as 25 Ω in order to attain wideband coupling of the LO pulse signal, as indicated in the design of the coupled-slotline magic-T circuit. The microstrip stub is terminated with a resistor, which is needed for the strobe pulse generator which requires a resistive load for the driving clock source, as well as for constructing a proper parallel tank circuit for the impulse generation. Figure 4.8 shows calculated results for the designed CSH with 50-Ω termination at the open end of the microstrip stub. Although the CSH was not measured, its operation is validated through good performance obtained for the sampler as will be seen later in Sect. 4.3.2. The insertion-loss result for the LO pulse signal in Fig. 4.8a shows a pass band from 1 to 5 GHz, which covers a significant portion of the frequency band of the pulse having 150 ps of FWHM. The return loss of the RF port and the isolation between the RF and LO ports displayed in Fig. 4.8b show good results up to 10 GHz. The transient simulation results in Fig. 4.8c show that the two split LO pulse signals (to two sampling diodes) are well matched and 180-deg out-of-phase without significant ringing. It should be noted that the first large side-lobe signal with opposite polarity to the main-lobe pulse, seen in Fig. 4.8c, does not cause any adverse effect to the sampling because it provides more reverse bias to the diode and effectively turns the diode off.

The layout for the sampling-bridge circuit shown in Fig. 4.7 was also optimized through simulations to minimize the effect of the layout on the performance of the RF transmission. Two radial slots are used at the sampling bridge to reduce the parasitic capacitances. Connection to the baseband circuit is implemented using via-holes and two underside microstrip lines. Close contact of the holding resistor to the sampling diode-bridge provides good isolation between the RF signal line and baseband connection.

An additional quarter-wavelength coupled-slotlines is used between the sampling-bridge and termination resistors to reduce possible ringing of the LO pulse signal at the sampling bridge. Without this transmission line, a positive reflected pulse signal occurring at the termination would broaden the sampling aperture time, leading to degradation of the bandwidth of the sampler. The quarter-wavelength line causes the reflected pulse signal to be aligned 180-deg out-of-phase with respect to the side-lobe (ringing) of the incident pulse signal at the sampling bridge, resulting in removal of the sidelobe of the pulse waveform. Effectively, the additional quarter-wavelength line provides better input matching at the sampling bridge for the LO pulse signal.

The use of CSH provides several advantages for the sampler. Firstly, by arranging the RF and LO ports on each side of the circuit, it is possible to make a good signal line termination taking advantage of the wide open slot. Secondly, the baseband connection can be implemented by the bottom microstrip lines, making easy integration with the baseband circuit. The designed sampling circuit does not require any wire connections or air bridges, hence reducing fabrication efforts and enabling low-cost mass production of the circuit, especially for hybrid MICs. A zero-ohm chip resistor is used in the CSH's coupled-slotlines at the RF port to simulate an air bridge needed for reflecting the LO pulse propagating toward that direction.

Fig. 4.8 Simulation results for the designed CSH. **a** S11: return loss at the LO-port, S31: insertion loss for the coupled LO pulse. **b** S44: return loss at the RF-port, S14: isolation between the RF and LO ports. **c** Pulse waveforms at the coupled-slotline output ports 2 and 3. Reference port numbers are shown in Fig. 4.7

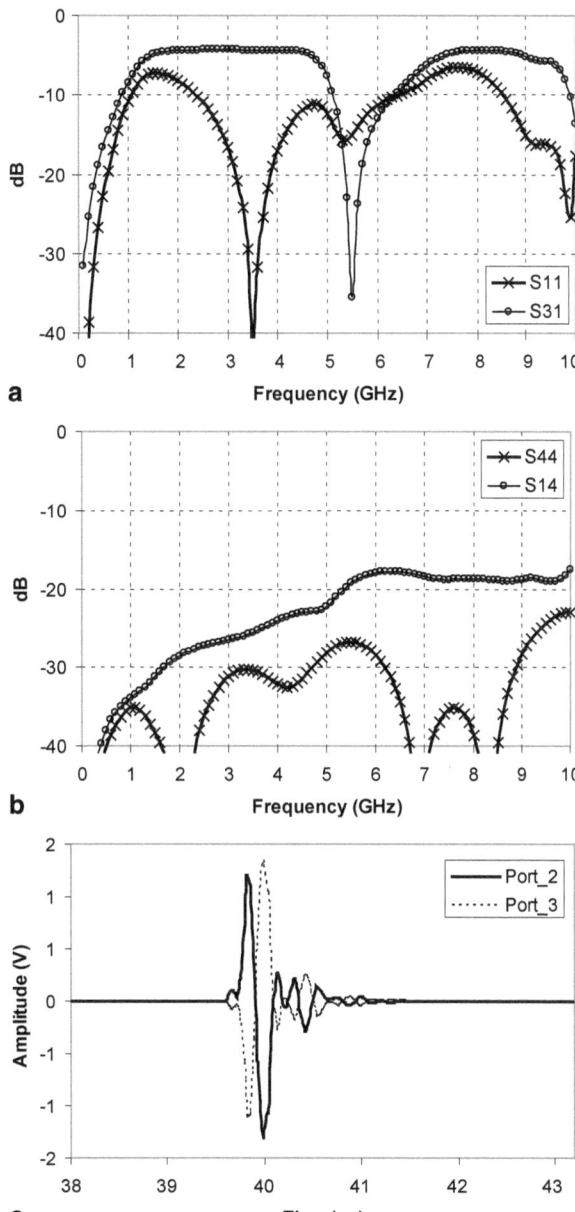

Fig. 4.9 Top and bottom
view of the CSH sampler
assembly. The sampling
bridge and the baseband
circuit are shown in the
top view **a** and the SRD
pulse sharpening circuit and
baseband connections are
shown in the bottom view **b**
of the assembly. The overall
dimension of the assembly is
3.3 in. × 2.0 in. × 0.6 in.

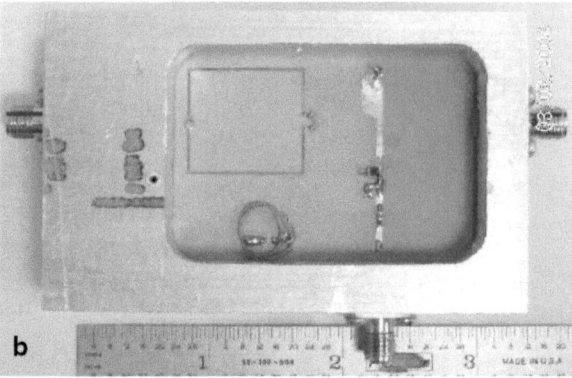

4.3.2   Fabrication and Performance of the CSH Sampler

The CSH sampling mixer was fabricated on RT/duroid 6010 substrate with a relative dielectric constant of 10.2, a thickness of 0.127 cm, and a loss tangent of 0.0023. The clock driving circuit of Fig. 4.3b was fabricated on a separate substrate for the convenience of measurement. Figure 4.9 shows the fabricated CSH sampling mixer assembled on a metal fixture without the clock driving circuit. The beam-lead Schottky-diode bridge used in the sampler is MSS-50244-B20, manufactured by Metelics Co. It has two diodes of high barrier needed for the sampler's high dynamic range. Each diode has a junction capacitance of 0.1 pF, junction resistance of 16 Ω, and series resistance of 11 Ω. The holding capacitors, shown in Fig. 4.7, are implemented by chip capacitors having low loss and high self-resonant frequency.

Figure 4.10 shows the measured strobe pulse waveforms at both the sampling bridge and the coupling area between the underside microstrip line and the coupled-slotlines. As indicated in Fig. 4.10, the negative side-lobe ringing, which can cause distortion on the sampled signal, is suppressed at the sampling bridge through the use of an additional quarter-wavelength coupled-slotlines mentioned in Sect. 4.3.1. The measured strobe pulse applied to the sampling bridge has a relatively small side-lobe ringing and similar waveform to the calculated pulse shown in Fig. 4.8c. The side-lobe ringing with opposite polarity to the main-lobe pulse does not cause

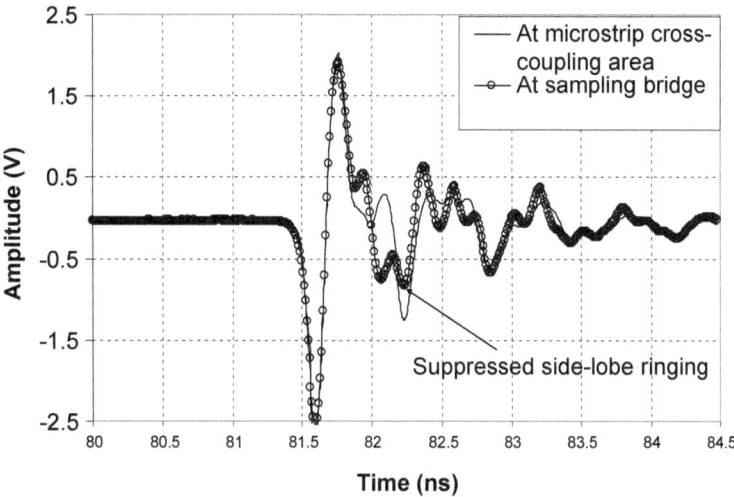

Fig. 4.10 Measured strobe pulse waveforms for the designed sampler

any problem for sampling as mentioned earlier. To find an optimum termination to the microstrip stub on the back side, the strobe pulse waveforms were measured for different values of the termination. The waveform shown in Fig. 4.10 is the best result obtained when a 25 Ω of termination was used.

Figure 4.11 shows the measured and simulated return losses at the RF port for the designed sampler. The measurement result shows a minimum return loss of 15 dB over the entire estimated 5.5-GHz bandwidth and a minimum return loss of 9 dB for a bandwidth up to 10 GHz. These results were obtained without the LO pulse signal. It is expected that the return loss without the LO pulse signal should resemble that with the LO signal because it is averaged out for a large time interval between the applied pulses. The measured return loss, therefore, corresponds to the diode turn-off condition. Accurate measurement for the return loss under the diode turn-on condition is not really necessary, whereas very cumbersome because it requires additional biasing networks to be attached. The measured return loss implies that the matching of the passive circuitry including the coaxial connection and CSH structure is well achieved. The simulation result, on the other hand, was done assuming the diodes were turned on, resulting in some difference with the measured data as noticed in Fig. 4.11. Nevertheless, the trend of the return loss curves matches each other reasonably well. The equivalent-circuit model shown in Fig. 4.6 was used for the sampling diodes in the simulation. It is deemed that the simulation result should provide a reasonably good estimate for the actual return loss when the sampling diodes are on.

Figure 4.12 shows the measured and calculated normalized (or relative) conversion loss of the CSH sampling mixer (without the baseband amplifier). The measured conversion loss is normalized to the minimum conversion loss of the sampler, which was measured as 4.5 dB. The calculated conversion loss is obtained using (4.2) and represents the intrinsic conversion loss of the sampler (normalized to the

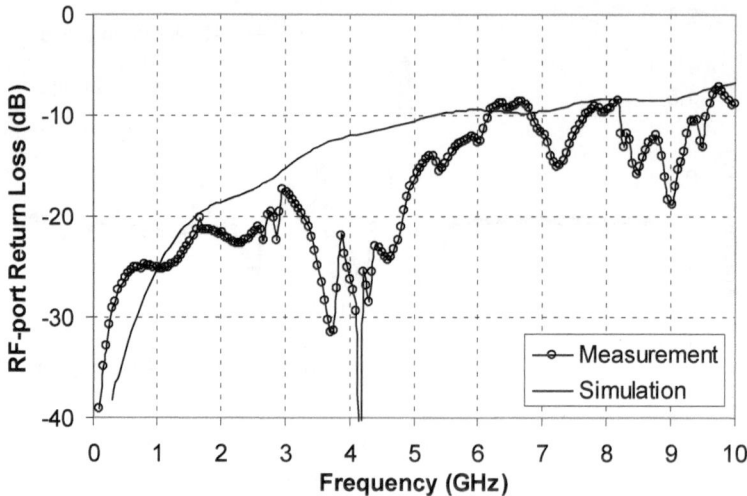

Fig. 4.11 Input return loss at the RF-port

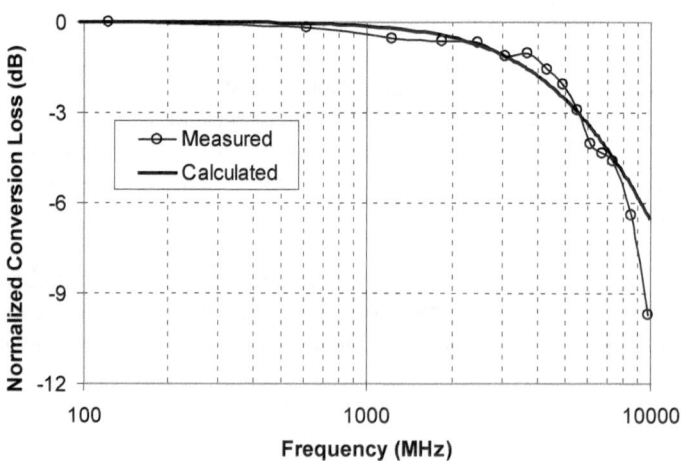

Fig. 4.12 Measured and calculated normalized conversion losses of the CSH sampler

minimum conversion loss) as described in sub-section 4.3.1. It is noted that the calculated intrinsic conversion loss matches very well with the actual measured normalized conversion loss, signifying the accuracy of (4.2) in predicting the normalized conversion loss and the operating bandwidth of the sampler. As can be seen in Fig. 4.12, the measured 3-dB bandwidth of the sampler is 5.5 GHz, which is in good agreement with the calculated one. With the measured 4.5-dB minimum conversion loss, the sampler exhibits a conversion loss from 4.5 to 7.5 dB for the RF signal from DC to 5.5 GHz and the baseband signal of 20 KHz. The sampler with

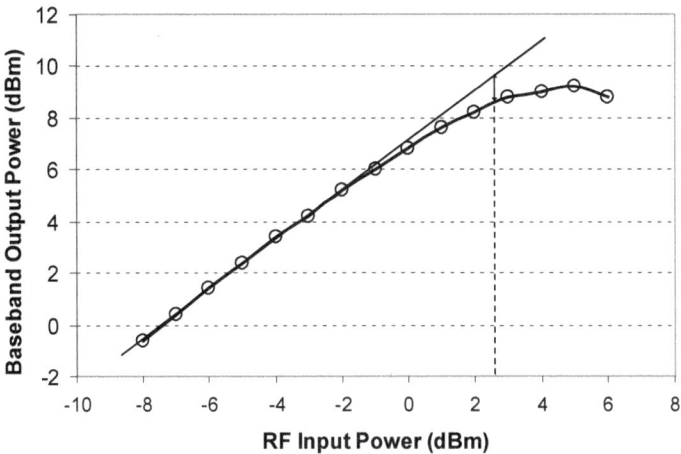

Fig. 4.13 Measured baseband output power of the CSH sampler

a 14-dB-gain (opamp) baseband amplifier shows a measured conversion gain from
6.5 to 9.5 dB over a 5.5-GHz RF bandwidth.

Figure 4.13 shows the measured baseband output power as a function of the RF
input power at a RF frequency of 3 GHz, which is near the center frequency of the
designed sampler. The measured 1-dB compression point is 2.5 dBm. To determine
the harmonic distortion in the baseband signal, the spurious signal, which is the
second harmonic of the down-converted signal, was measured using a spectrum
analyzer. For the RF signal of 3 GHz and 0 dBm, the spurious signal level is as low
as 20 dBc.

The sensitivity of the sampler was obtained by measuring the noise output power
of the baseband signal using a spectrum analyzer. The measured output noise power
is -33 dBm for a 3-GHz, -9.9-dBm RF input signal and 2 dBm of the baseband
signal output. Using the sensitivity of the spectrum analyzer and the baseband filter
bandwidth (1.5 MHz) of the sampler, the sampler's sensitivity, defined as 8-dB tangential sensitivity, was determined as −47 dBm. The dynamic range of the sampler
was measured to be more than 50 dB. A summary of the sampler's performance is
shown in Table 4.2.

4.4   UWB Receiver

4.4.1   Design of the UWB Receiver

As discussed in Sect. 4.1, the UWB receiver used in the UWB system presented
in this book is based on the synchronous sampling technique using two oscillators
of slightly different frequencies. For the synchronous sampling, the two reference

Table 4.2. Summary for the performance of the designed CSH sampler

Conversion loss (−) and gain (+)	−7.5 to −4.5 dB (without amplifier)
	6.5 to 9.5 dB (with amplifier)
Baseband amplifier gain	14 dB
RF input bandwidth	5.5 GHz
LO clock source PRF	12 MHz
Spurious	20 dBc typical
1-dB compression	2.5 dBm
8-dB sensitivity	−47 dBm
Dynamic range	>50 dB
RF VSWR	1.3:1

clock oscillators should provide proper pulse repetition rates for the LO strobe pulse signal and the transmitted signal (and hence the RF input signal to the receiver). From the system level perspective, the design of the UWB receiver necessitates the determination of the PRFs of the two clock oscillators required for proper synchronous sampling.

Let us assume that the PRF of the LO strobe pulse is f_o and that of the RF input signal is $f_R = Nf_o + \Delta f$, where N is the integer representing harmonics and Δf is the frequency deviation. This relationship between the PRFs of the LO and RF pulse signals is the familiar form in the harmonic mixing theory and, in our particular case, $N = 1$. Referring to Fig. 4.1, the periods for the RwF and LO pulse signals, T_R and T_o, respectively, can be written as

$$T_R = \frac{1}{f_R} = \frac{1}{f_o + \Delta f} \tag{4.4}$$

$$T_o = \frac{1}{f_o} = T_R + \Delta t \tag{4.5}$$

where Δt is the (equivalent) sampling time interval. Using (4.4) and (4.5), we can obtain the deviation frequency Δf in terms of Δt and f_o as

$$\Delta f = \frac{\Delta t f_o^2}{1 - \Delta t f_o} \tag{4.6}$$

Note that $f_o \approx f_R$ as considered here and f_R is needed to determine a proper value for Δf.

The transmitter's clock oscillator frequency determines the PRF of the transmitting pulse, which is the same as that of the RF input signal, f_R, and depends on the required observation time interval for a particular application and the scanning speed of the system. The required observation time interval is dependent on the velocity of the wave propagating in the target structure and the maximum desired detection range. For short-range sensing, this observation time is very small—within a few tens of ns. The scanning speed of the system depends on the platform carrying it and practically dictates the PRF. For vehicular platforms, this speed may be more than 70 miles per hour. As an example, we consider a UWB system on a moving vehicle that senses a target represented by a pavement consisting of asphalt, base, and sub-base whose parameters are given in Table 2.1 and 2.2 of Chap. 2. The required observation time interval can be calculated as greater than 10 ns. Considering the required observation time interval and the scanning speed of the UWB system, the PRF of the transmitting pulse can be estimated as $f_R = 10$ MHz. Using this value together the fact that f_o is very close to f_R and selecting $\Delta t = 1$ ps, the required frequency deviation is calculated as 100 Hz from (4.6).

One interesting question is how many sampling points (or samples) would be produced for a single pulse-repetition period of the RF input signal. In other words, what is the value M for a single period T_R in Fig. 4.1? It is known that Δf is the PRF of the down-converted signal according to general harmonic mixing theory. Therefore, a single pulse-repetition period of the down-converted signal can be obtained as

$$\frac{1}{\Delta f} = M T_R \tag{4.7}$$

where M is, making use of (4.4),

$$M = \frac{f_o + \Delta f}{\Delta f} \tag{4.8}$$

Using $f_o = 10$ MHz, $\Delta f = 100$ Hz, M = 100K samples can be calculated from (4.8). The calculated number of sampling points, 100K, seems too big for a single pulse-repetition period, which is a single scan data at one location in the target structure. However, the actual number of the samples per single scan is varied by the ADC (analog-to-digital) processing upon the down-converted signal. For instance, if the sampling rate of the ADC, f_s, is the same as $f_o = 10$ MHz, the sampled time interval on the RF input pulse is $\Delta t = 1$ ps. Assuming a 450-ps-duration monocycle pulse is used as the transmitting pulse, 450 sampling points would be reproduced in the baseband for a time interval corresponding to the pulse duration, which is obviously an excessively large number of samples. A rule of thumb in the sampling process is that 10 samples for a pulse-duration of a pulse signal can provide good resolution in the ADC process. If the sampling rate of the ADC is 45 times less than $f_o = 10$ MHz, that is $f_s = 220$ KHz, we can make 10 samples per pulse-duration in this example. For the UWB system presented in this book, the minimum RF pulse duration is 450 ps, which is the minimum pulse duration of the UWB transmitter designed in Chap. 3, so the sample rate $f_s = 220$ KHz is also appropriate. Using this lower sampling rate, the cost for an ADC can be reduced significantly, and the actual number of the sampling points per single scan is reduced to 2.2K samples for 100-Hz baseband PRF. Table 4.3 summarizes the design parameters derived for implementation of the UWB receiver used in the UWB system.

Table 4.3 Design parameters for implementing
the UWB receiver for
a transmitting pulse of
450-ps duration

Reference clock frequency (f_o)	10 MHz
Equivalent sampling time interval (Δt)	1 ps
Deviation frequency (Δf)	100 Hz
Sampling rate of ADC (f_s)	220 KHz
Samples/scan	2.2K

Selecting two appropriate reference clock oscillators for the transmitter and the receiver is vital for implementation of the UWB receiver. First, we need to select the reference clock oscillators according to the design parameters of $f_o = 10$ MHz and $\Delta f = 100$ Hz. Second, the oscillator output frequency should be very stable with respect to temperature variation. Third, the oscillators should have a small jitter (or phase noise) because a large jitter may cause additional distortion and conversion loss on the down-converted signal due to the sampling timing error.

Temperature controlled crystal oscillators (TCXOs) producing 5V HCMOS logic output are usually unable to meet the foregoing requirements at low cost. One way to achieve low-cost implementation is to use voltage controlled TCXOs (VC-TCXOs). Most of VC-TCXO manufacturers provide about 10 ppm of frequency adjustment through voltage control, and 100-Hz frequency variation is possible. An even larger frequency adjustment range can be achieved by using additional mechanical trimmer of a VC-TCXO. Two VC-TCXOs are finally selected for our design, which are the GTXO-536V model manufactured by Golledge Electronics because it has very low phase noise (−135 dBc/Hz @ 1 KHz) and is a low-cost product.

4.4.2   Fabrication and Performance of the UWB Receiver

As indicated in Fig. 4.2, the UWB receiver used in the UWB system presented in this book is essentially the integrated sampling mixer which includes the sampling mixer, the strobe pulse generator, reference clock generator, and baseband circuit. In the actual UWB receiver, a VC-TCXO (model GTXO-536V mentioned earlier) is used as the reference clock generator (specifically in the clock driving circuit of the strobe pulse generator of the receiver). It is noted that another identical VC-TCXO is also used in the clock driving circuit of the transmitter as the reference clock. Figure 4.14 shows a photograph of the UWB receiver fabricated on a single PCB using the same substrate as the one used for the CSH sampling mixer described in Sect. 4.3.2.

The performance of the UWB receiver is measured in terms of the quality of its reproduced baseband signal in comparison with that reproduced by a 12-GHz commercial sampling oscilloscope. Figure 4.15 shows the measurement system used to measure both the down-converted signals by the UWB receiver and the sampling oscilloscope. In Fig. 4.15, Path 1 represents the measurement setup to measure the transmitter's output signal down-converted by the sampling oscilloscope through its own down-conversion function. Path 2 is the setup to measure the same transmitter's output signal down-converted by the UWB receiver and measured by the sampling oscilloscope. The designed UWB transmitter described in Sect. 3 is used

Fig. 4.14 Top view of the fabricated UWB receiver circuit

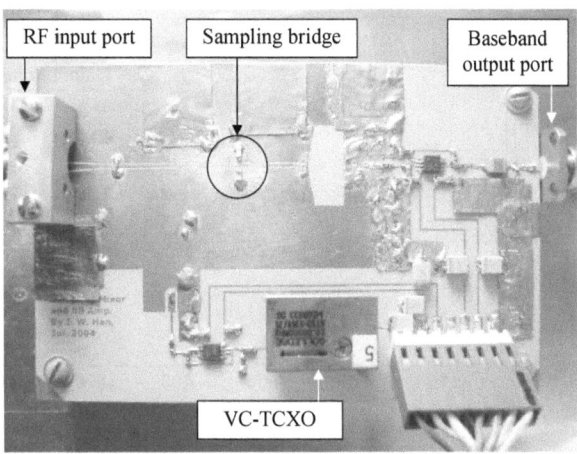

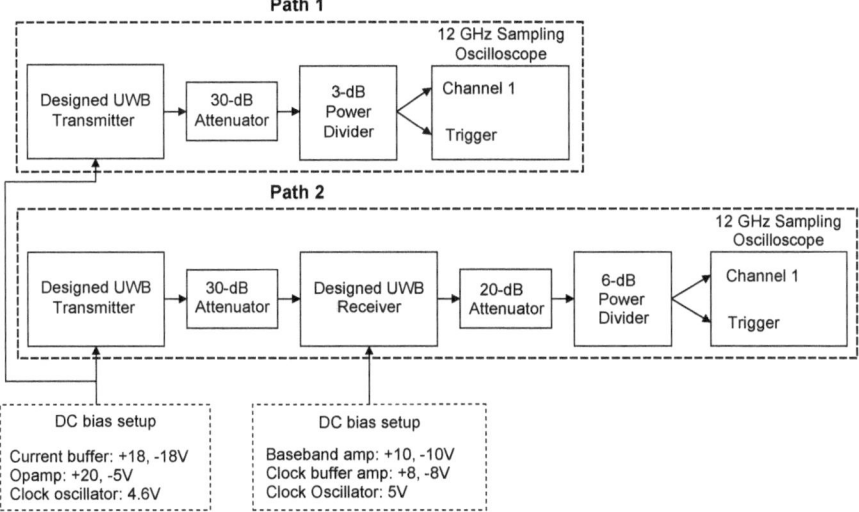

Fig. 4.15 Setup for measuring the signals down-converted by the UWB receiver and the sampling scope

here to test the performance of the UWB receiver. Use of the designed transmitter in these experiments also demonstrates the workability and performance of the transmitter as well as its implementation for UWB systems.

The measurements of the down-converted signals for Path-1 and Path-2 setup were recorded in the sampling oscilloscope. The transmitter's output signal used for these measurements is a monocycle pulse with duration of 450 ps, which is the minimum pulse duration of the designed UWB transmitter. This minimum pulse duration is used to test whether the bandwidth of the designed UWB receiver can work up to the maximum bandwidth of the tunable transmitting pulse produced by the

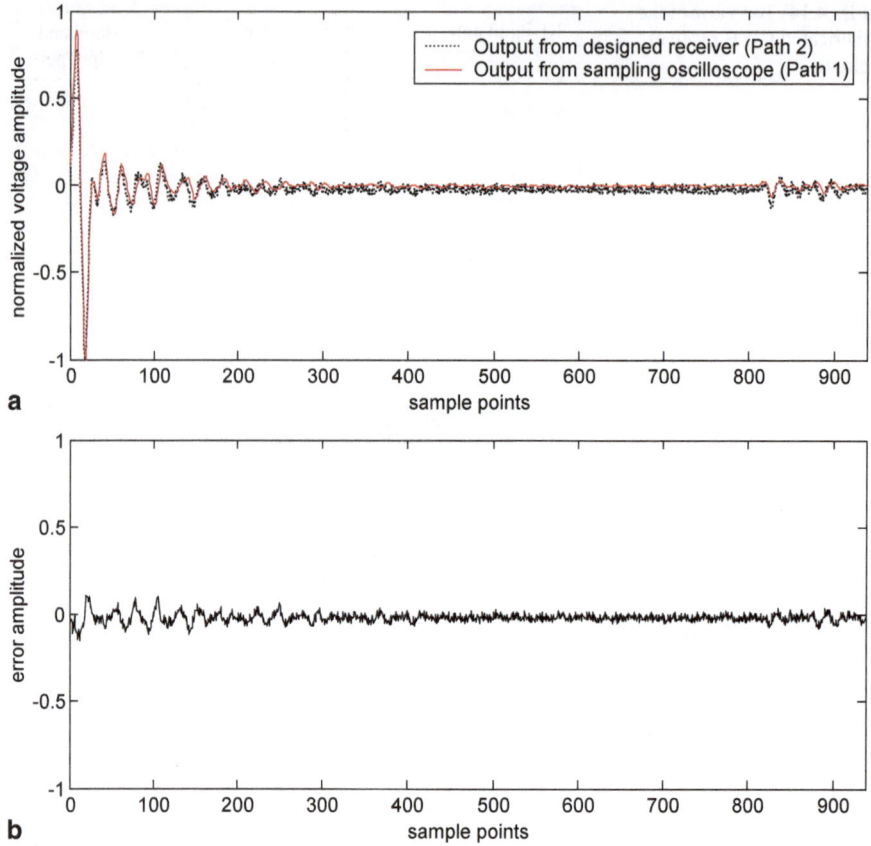

Fig. 4.16 a Measured waveforms of pulse signals down-converted by the designed UWB receiver and the sampling scope. **b** Errors between the two measured waveforms

transmitter, which is necessary for the UWB system to work properly. Figure 4.16 shows the signals down-converted by the designed UWB receiver and by the 12-GHz sampling oscilloscope, and the errors between them. As can be seen, the two down-converted signals match very well. The root mean square (RMS) error between these signals is only 0.035, signifying that the difference between the two down-converted signals is only 3.5 % of the signal's peak amplitude. These results demonstrate that the receiver can reconstruct the incoming pulse signals with negligible error.

4.5   UWB LNA

UBW LNA is important for UWB systems in applications and/or operations that result in relatively weak input signals to the UWB receivers of the systems. Besides the usual requirements of high gain, low noise figure, and high linearity across a desired ultra-wide band, a LNA used for UWB systems must possess high signal-fidelity characteristics so that it can amplify a UWB signal with minimum distortion and hence preserve well the shape of the input UWB signal waveform.

Distributed amplifiers are an attractive candidate for UWB systems—not only because of their inherent extremely wide bandwidth, but also because of their excellent phase linearity of the transmission response which leads directly to minimum signal distortion. The superior phase linearity is due to the exclusively use of transmission lines in the design of the amplifiers. Well-designed transmission lines have small velocity variation versus frequency, leading to highly linear transmission phase, as well as low loss and small loss variation with frequency—all leading to minimum distortion for signals traversing the transmission lines. In this section, we will present a brief design and the time-domain performance of a distributed amplifier to demonstrate its possible use as an UWB LNA for UWB receivers. It is noted that, as an UWB system transmits and receives (non-sinusoidal) pulse signals, the most accurate design and characterization of components used in a UWB system should be carried out in the time domain. This approach was particularly carried out for the UWB LNA presented here. The design and test of the other components for the UWB system described in this book such as the sampling mixer were not conducted completely in the rime domain primarily for convenience. The distributed amplifier is designed using radio-frequency integrated-circuit technology implemented on a 0.25-μm complementary metal–oxide–semiconductor (CMOS) process. Detailed design of distributed amplifiers can be found in [36].

Figure 4.17 shows the schematic and photograph of the designed UWB distributed amplifier. Six transistors and finite-ground coplanar waveguide (CPW) are used in the amplifier. The (CPW) transmission line connecting the gates and that connecting the drains are periodically loaded with the transistor's gate-source and drain-source capacitance, respectively. Together, they form artificial transmission lines at the gates (gate line) and drains (drain line) of the transistors. The gate line and drain line are terminated with resistors having resistances of 50 Ohms, which are the artificial transmission lines' characteristic impedances. On-chip capacitor C blocks the DC path from the bias to the terminal resistors. On-chip inductors (L), acting as RF chokes, feed the DC voltages and currents to the circuit, while stopping the RF signals from going through it. Finite-ground CPW is used to allow easy connection of the sources of the transistors to the ground while maintaining a compact size for the amplifier. CPW also facilitates wider central strip for high characteristic impedance than its microstrip line counterpart, thus lowering the conductor loss. Low loss is needed to maintain well-behave artificial transmission lines, which are critical for realizing distributed amplifiers. The topmost metal layer is chosen for the CPW to obtain low loss due to its thickest metallization and farthest distance from the silicon substrate.

Figure 4.18 shows the calculated and measured transmission-phase responses of the UWB distributed amplifier. The results show that the amplifier exhibits a very linear transmission-phase response across an extremely wide frequency range. Linear phase response, and hence constant velocity, across a wide frequency range is very critical for UWB systems.

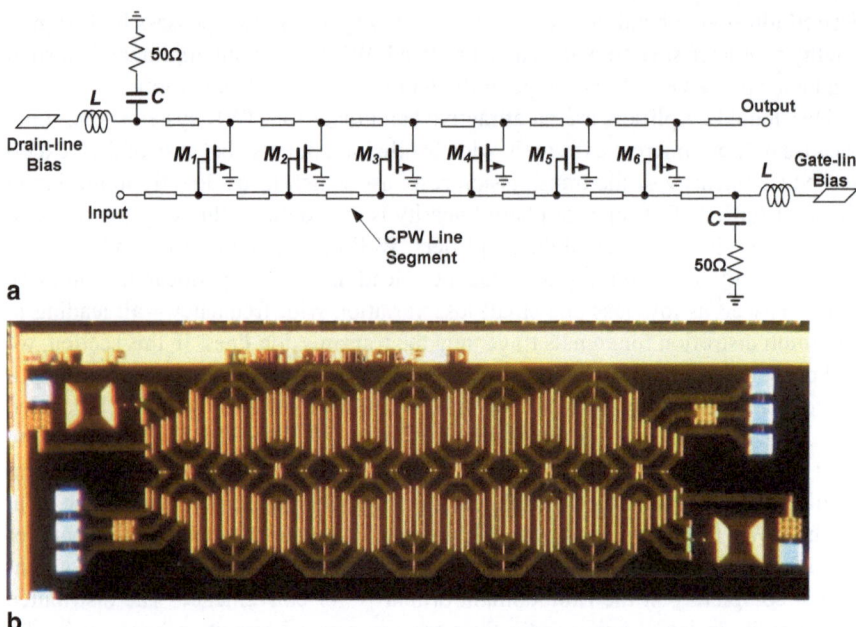

a

b

Fig. 4.17 Schematic **a** and photograph **b** of the UWB distributed amplifier

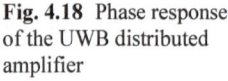

Fig. 4.18 Phase response
of the UWB distributed
amplifier

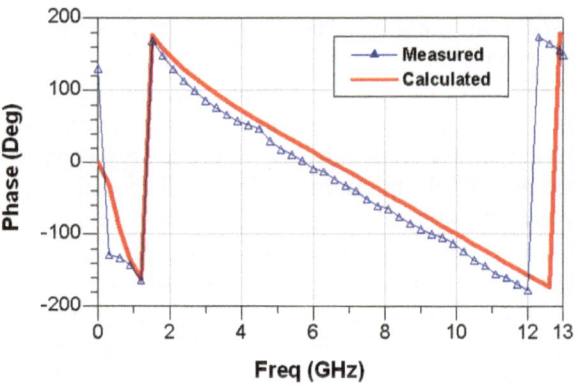

To assess the performance of the amplifier and its suitability for UWB applications, both simulation and measurement have been conducted in the time domain. Figure 4.19 shows the simulated waveform of the output signal for an input impulse of 100-ps 3-dB pulse-width. The output waveform is distorted as expected due to the fact that the amplifier does not pass the low-frequency and DC components of the input impulse, which shows indeed a drawback of using impulse signals as discussed in Sect. 2.3 of Chap. 2. Figure 4.20 shows the simulated output waveform for an input monocycle pulse with 200-ps 3-dB pulse-width. Figures 4.21 and 4.22

Fig. 4.19 Time domain simulation result with impulse input

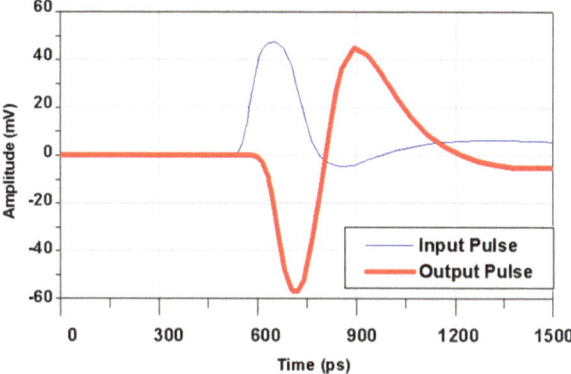

Fig. 4.20 Time domain simulation result with monocycle pulse input

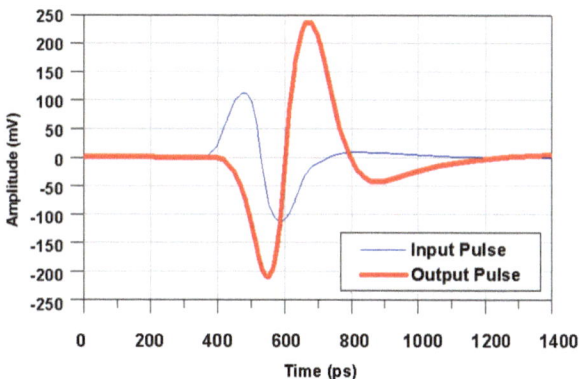

Fig. 4.21 Measured output pulse with 100-ps impulse input

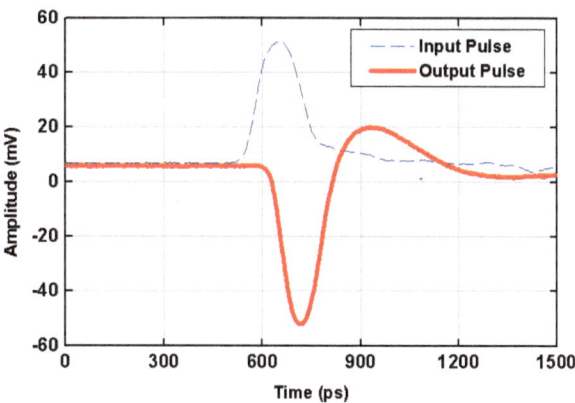

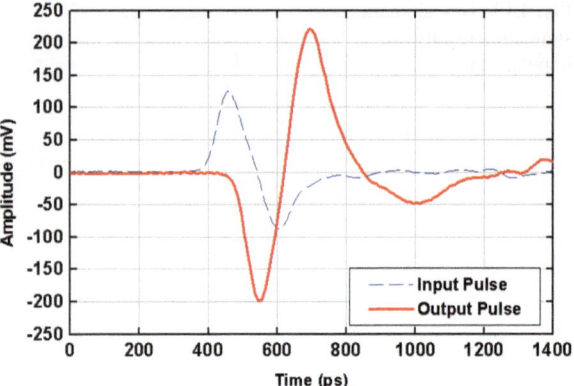

Fig. 4.22 Measured output pulse with 200-ps monocycle pulse input

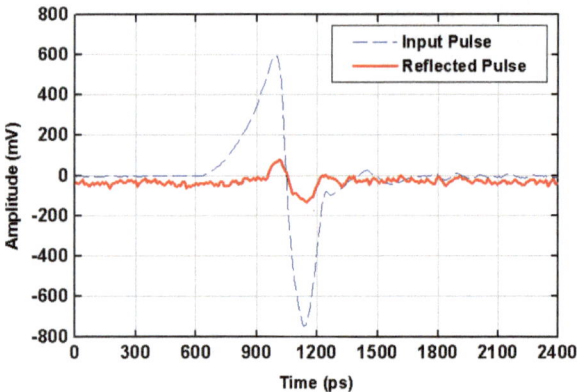

Fig. 4.23 Measured reflected signal at the input port for 200-ps monocycle pulse input signal

show the measured output signals for a 100-ps impulse and 200-ps monocycle pulse input, respectively. The measured gain is around 6 dB for the monocycle pulse input. The gain is lower for the impulse input signal because the amplifier does not pass the low-frequency and DC components as indicated earlier. As can be seen, the output waveforms match closely those at the input with similar pulse duration and very little distortion. This faithful reproduction of the signal waveforms demonstrates that the developed UWB amplifier fulfills one of the most critical requirements for UWB systems used for (time-domain) UWB applications, namely retaining the shape and duration of a transmitted pulse. This high signal fidelity is a critical requirement for UWB systems, as mentioned early, as the UWB pulse signal carrying information must be transmitted with minimum distortion.

A broadband directional coupler was also used to measure the reflection at the input port of the UWB amplifier. Figs. 4.23 and 4.24 show the measured reflected signals at the input port for the 100-ps impulse and 200-ps monocycle pulse input, respectively. More than 13-dB return loss is obtained. The input pulses displayed in the figures are the signals reflected from an open circuit. Both the input and reflected signals were measured at the same coupling port of the directional coupler.

Fig. 4.24 Measured reflected signal at the input port for 100-ps impulse input signal

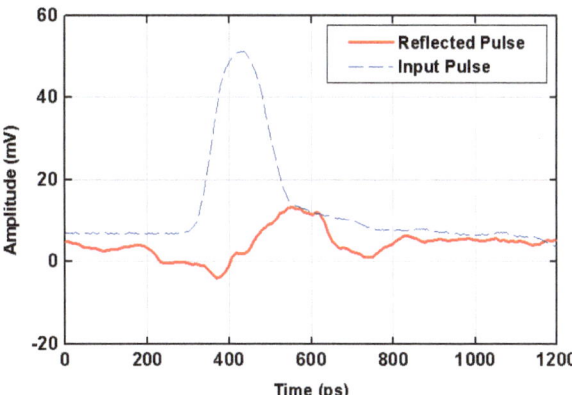

4.6 Summary

The design of an UWB receiver for UWB systems has been described. The receiver consists of a sampling mixer, strobe pulse generator, reference clock generator, and baseband circuit. It is realized in a single circuit board using microstrip line, CPW, slot line, and coupled slot lines, and is compact and low cost. The receiver achieves a conversion loss of 4.5–7.5 dB (without baseband amplifier) and conversion gain from 6.5–9.5 dB (with amplifier) across a 5.5-GHz RF bandwidth, dynamic range of more than 50 dB, and low harmonic distortion in the baseband output signal. Its performance in down-converting signals is comparable to a commercial sampling scope, yet with much significantly smaller size and lower cost. The design of the strobe pulse generators for sampling applications requiring low PRF was also presented. Detailed design of the sampling mixer was described. Moreover, a distributed amplifier suitable for UWB receivers was discussed. The designed amplifier was characterized completely in the time domain under non-sinusoidal signal operation necessary for UWB applications. It has a highly linear transmission phase and produces amplified signals which resemble faithfully the input UWB pulses, demonstrating its suitability for UWB applications.

References

1. Daniels, D.J.: Surface Penetrating Radar. IEEE Press, London (1996)
2. Taylor, J.D.: Introduction to Ultra-Wideband Radar Systems. CRC Press, Boca Raton (1995)
3. Abuasaker, S., Kompa, G.: A high sensitive receiver for baseband pulse microwave radar sensor using hybrid technology. IEEE Radar Conference Proceedings, pp. 121–124. (2002)
4. Weingarten, K.J., Rodwell, M.J.W., Bloom, D.M.: Picosecond optical sampling of GaAs integrated circuits. IEEE J. Quantum Electron. **24**(2):198–220 (February 1988)
5. Kamegawa, M., Giboney, K., Karin, J., Allen, S., Case, M., Yu, R., Rodwell, M.J.W., Bowers, J.E.: Picosecond GaAs monolithic optoelectronic sampling circuit. IEEE Photonics Tech. Lett. **3**(6):567–569 (June 1991)

6. Akos, D.M., Tsui, J.B.Y.: Design and implementation of a direct digitization GPS receiver front end. IEEE Trans. Microw. Theory Tech. **44**(12):2334–2339 (December 1996)

7. Taylor, J.D.: Ultra-Wideband Radar Technology. CRC Press, Boca Raton (2001)

8. Grove, W.M.: Sampling for oscilloscopes and other RF systems: DC through X-band. IEEE Trans. Microw. Theory Tech. MTT-**14**(12):629–635 (December 1966)

9. Merkelo, J., Hall, R.D.: Broad-band thin-film signal sampler. IEEE J. Solid-State Circuits. SC-**7**(1):50–54 (February 1972)

10. Akers, N.P.: RF sampling gates. A brief reviw. IEE Proc. **133**(1):45–49 (January 1986)

11. Bologlu, A.: A 26.5-GHz automatic frequency counter with enhanced dynamic range. Hewlett-Packard J., 20–22 (April 1980)

12. Gilchrist, B.E., Fildes, R D., Galli, J.G.: The use of sampling techniques for miniaturized microwave synthesis applications. In 1982 IEEE MTT-S International Microwave Symposium Digest, pp. 431–433 (1982)

13. Madani, K., Aichison, C.S.: A 20GHz Microwave Sampler. IEEE Trans. Microw. Theory Tech. **40**(10):1960–1963 (October 1992)

14. Moore, S.E., Gilchrist, B.E., Galli, J.G. Microwave sampling effective for ultrabroadband frequency conversion, MSN & CT, pp. 113–126 (February 1986)

15. Gibson, S.R.: Gallium arsenide lowers cost and improves performance of microwave counters. Hewlett-Packard J. **1986**, 4–10 (February 1986)

16. Shakouri, M.S., Black, A., Auld, B.A., Bloom, D. M.: 500 GHz GaAs MMIC sampling wafer probe. Electron. Lett. **29**(6):557-558 (March 1993)

17. Rodwell, M.J.W., Kamegawa, M., Yu, R., Case, M., Carman, E., Giboney, K.S.: GaAs nonlinear transmission lines for picosecond pulse generation and millimeter-wave sampling. IEEE Trans. Microw. Theory Tech. **39**(7):1194–1204 (July 1991)

18. Konishi, Y., Kamegawa, M., Case, M., Yu, R., Allen, S.T., Rodwell, M.J.W.: A broadband free-space millimeter-wave vector transmission measurement system. IEEE Trans. Microw. Theory Tech. **42**(7):1131–1139 (July 1994)

19. Miura, A.: Monolithic sampling head IC. IEEE Trans. Microw. Theory Tech. **38**(12):1980–1985 (December 1990)

20. Marsland, R.A., Valdivia, V., Madden, C.J., Rodwell, M.J.W., Bloom, D.M.: 130 GHz GaAs monolithic integrated circuit sampling head. Appl. Phys. Lett. **55**(6):592–594 (August 1989)

21. Abele, P., Birk, M., Behammer, D., Kibbel, H., Trasser, A., Maier, P., Schad, K.-B., Sönmez, E., Schumacher, H.: Sampling circuit on silicon substrate for frequencies beyond 50 GHz. IEEE MTT-S International Microwave Symposium Digest, pp. 1681–1684 (2002)

22. Pärssinen, A., Magoon, R., Long, S.I., Porra, V.: A 2-GHz subharmonic sampler for signal downconversion. IEEE Trans. Microw. Theory Tech. **45**(12):2344–2351 (December 1997)

23. Lee, J.S., Nguyen, Cam.: A low-cost uniplanar sampling down-converter with internal local oscillator, pulse generator, and IF amplifier. IEEE Trans. Microw. Theory Tech. **49**(2):390–392 (February 2001)

24. Han, J., Nguyen, C.: Integrated balanced sampling circuit for ultra-wideband communications and radar systems. IEEE Microw. Wireless Compon. Lett. **14**(10):460–462 (October 2004)

25. Han, J.W., Nguyen, C.: Coupled-Slotline-Hybrid Sampling Mixer Integrated with Step-Recovery-Diode Pulse Generator for UWB Applications. IEEE Trans. Microw. Theory Tech. MTT-**53**(6):1875–1882 (June 2005)

26. Lee, J.S., Nguyen, C., Scullion, T.: Impulse ground penetrating radar for nondestructive evaluation of pavements. IEEE MTT-S International Microwave Symposium Digest, pp. 1361–1363 (2002)

27. Williams, D.F., Remley, K.A.: Analytic sampling-circuit model. IEEE Trans. Microw. Theory Tech. **49**, 1013–1019 (June 2001)

28. Remley, K.A.: Realistic sampling-circuit model for a nose-to-nose simulation. IEEE MTT-S International Microwave Symposium Digest, pp. 1473–1476 (2000)

29. Aikawa, M., Ogawa, H.: A new MIC magic-T using coupled slot lines. IEEE Trans. Microw. Theory Tech. MTT-**28**(6):523–528 (June 1980)

30. Hamilton, S., Hall, R.: Shunt-mode harmonic generation using step recovery diodes. Microw. J. 69–78 (April 1967)
31. Moll, J.L., Hamilton, S.: Physical modeling of the step recovery diode for pulse and harmonic generation circuits. Proc. IEEE. **57**(7):1250–1259 (July 1969)
32. Goldman, S.: Computer aids design of impulse multipliers. Microwaves & RF, pp. 101–128 (October 1983)
33. Zhang, J., Räisänen, V.: Computer-aided design of step recovery diode frequency multipliers. IEEE Trans. Microw. Theory Tech. **44**(12):2612–2616 (December 1996)
34. Han, J., Nguyen, Cam.: A new ultra-wideband, ultra-short monocycle pulse generator with reduced ringing. IEEE Microw. Wireless Compon. Lett. **12**(6):206–208 (June 2002)
35. Advanced Design System.: Agilent Technologies, Inc., Santa Clara, CA, U.S.A
36. Nguyen, C.: Radio-Frequency Integrated-Circuit Engineering. Wiley, New York (2014)

References

Chapter 5
UWB Antenna Design

5.1 Introduction

The antenna for the transmission and reception of UWB pulse signals is another
important component of UWB systems. UWB impulse circuits and systems involve
propagations of (non-sinusoidal) pulse signals, not multiple discrete frequency
components in CW mode, through circuits, antennas, and air or other media. UWB
systems typically transmit and receive information using millions of narrow pulses
each second across an ultra-wide band spectrum. Effectively, UWB systems, and
hence UWB antennas, transmit and receive all frequency components across extremely wide bandwidths simultaneously, not consecutively. Therefore, not only
that the UWB antenna needs to work over an ultra-wide frequency range covering
the bandwidths of UWB signals, but it also needs to radiate or receive these UWB
signals with no distortion (ideally) or as small distortion as possible. This results in
another constraint for the UWB antenna design; in which linear phase needs to be
preserved across the interested band so that the antenna can reproduce faithfully the
waveforms of the input UWB signals. This high signal fidelity is a critical requirement for antennas used in UWB systems, as the UWB signals carrying information
must be transmitted and received with minimum distortion. Similar to other antenna
types, the antenna gain for pulse signals is also an important factor for UWB systems. High-gain UWB antennas enable UWB signal transmission with low loss and
hence high signal-transmission efficiency. For UWB signal reception, high-gain
UWB antennas lead to enhanced signal detection. Another important criterion for
UWB antennas is their matching. As an UWB antenna transmits and receives pulse
signals, not only the reflection at the antenna's input needs to be reduced, but also
those along the antenna structure need to be minimized as well, as these reflections
can degrade the antenna performance and hence UWB systems. All these requirements make the design of UWB antennas more challenging than that of typical
antennas supporting CW signals, even for the same operating frequency range.

Various antennas can be used for pulse signals—for instance, transverse electromagnetic (TEM) horn, bow-tie, log-periodic, spiral, and conical antennas [1–8].
Among them, the TEM horn antenna employing parallel-plate transmission line is
perhaps the most commonly used antenna for UWB systems due to its relatively

C. Nguyen, J. Han, *Time-Domain Ultra-Wideband Radar, Sensor and Components*,
SpringerBriefs in Electrical and Computer Engineering,
DOI 10.1007/978-1-4614-9578-9_5, © Springer International Publishing Switzerland 2014

high gain, wide band, unidirectional radiation, no dispersion (ideally), and small loss through the transmission line constituting the antenna. These properties are particularly important for UWB applications in order to overcome the high loss in propagating distances and/or target structures in which the UWB signals traverse, as well as to preserve the shape of the transmitted and received waveforms. Many efforts have been spent to study and develop the TEM horn antennas. The TEM horn antennas, however, prohibit a direct integration of the antennas with microwave integrated circuits (MICs). In addition, they require a balun at the input, thereby limiting the operating bandwidth, especially at high microwave frequencies. Moreover, the TEM horn antennas, when used as transmit and receive elements, suffer from an inherently strong coupling between them when placed next to each other.

A modified version of the TEM horn antenna, namely quasi-horn antenna, was developed [7] and used successfully for several UWB subsurface penetrating radars, e.g., [9, 10]. The quasi-horn antenna employs non-uniform quasi-TEM transmission lines, such as microstrip transmission line, coplanar waveguide (CPW), or coplanar strips, and has extremely broad bandwidth of multiple decades, relatively high gain, unidirectional radiation, little dispersion, and excellent linear phase. Moreover, with proper size and shape, the quasi-horn antenna can achieve a relatively flat gain response over an extremely wide bandwidth. It has several advantages as compared to the conventional TEM horn. First, its physical size is smaller than the TEM horn. Second, it employs such transmission lines that are more conveniently fabricated and easily integrated with MICs. Third, when an unbalanced transmission line such as microstrip line is used, the antenna requires no transition or balun at its input, leading to possibly wider bandwidths and higher operating frequencies. Lastly, when two identical quasi-horn antennas realized using microstrip line are used as transmit and receive antennas in a bi-static UWB system, there is an inherent isolation between the antennas, even when they are placed next to each other and no absorbing material is used, due to the common ground plane between them. This is especially important for system operation since the cross-coupled signals, occurred between two non-isolated antennas, would distort the desired received signals. Another UWB antenna having a uniplanar structure, which can be easily integrated with planar integrated-circuit transmitters or receivers, was also developed [11]. This UWB uniplanar antenna employs non-uniform slot lines and coplanar waveguide (CPW), and can operate over a very wide bandwidth with omni-directional radiation and high signal fidelity.

In this chapter, we present the design of two UWB antennas including the design approach, fabrication and measurement: one is the UWB microstrip quasi-horn antenna and another is the UWB uniplanar antenna.

The microstrip quasi-horn antenna was designed for transmission and reception of monocycle pulse signals with pulse duration in the range of 400 to 1200 ps, which corresponds to the output pulse duration of the designed UWB transmitter described in Chap. 3. The measured performance of the designed microstrip quasi-horn antenna shows 2:1 input VSWR from 0.3 to a more than 6 GHz. The measured time-domain return loss of the antenna, based on pulse transmission and reflection,

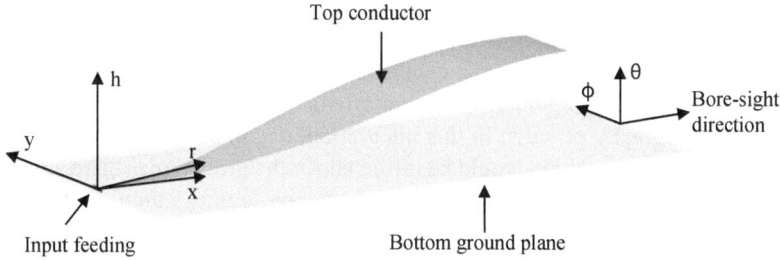

Fig. 5.1 General configuration of the quasi-horn antenna with non-uniform microstrip line as the transmission line

is about 13, which represents the actual return loss of the antenna for (impulse) UWB systems. The measured antenna gain is more than 12 dBi at frequencies higher than 2.5 GHz. The time-domain measurement results show that the designed antenna can transmit and receive 400 to 1200 ps monocycle pulses effectively with small waveform distortion and ringing.

The designed UWB uniplanar antenna can transmit and receive faithfully the impulse and monocycle pulse signals having varying durations of 100–300 ps and 140–350 ps, respectively. The measured impulse and monocycle-pulse transmission results demonstrate the workability of the antenna for UWB applications. In the frequency domain, the designed UWB uniplanar antenna exhibits more than 11-dB return loss from 3 to 12 GHz while, in the time domain, it shows better than 18-dB measured return loss.

5.2   UWB Quasi-Horn Antenna

5.2.1   Design of UWB Quasi-Horn Antenna

Figure 5.1 shows the basic structure of the quasi-horn antenna using non-uniform microstrip line. It is noted that microstrip line is used as an example here to illustrate the design without loss of generality; other transmission lines can also be used with proper arrangement. The microstrip quasi-horn antenna consists of a conductor of varying width on top of a grounded dielectric substrate which constitutes a non-uniform microstrip line. The dielectric substrate can be shaped in any particular fashion, and the conductor's (width) profile depends on the contour of the dielectric substrate. The antenna is basically an end-fire traveling-wave antenna employing non-uniform transmission line. If air is used as the dielectric medium and another identical conductor is placed symmetrically with respect to the conductor without the ground plane, then the structure can be considered as a TEM horn antenna consisting of two conductors of the same width and separated by twice the distance. These corresponding microstrip quasi-horn and TEM horn antennas are electrically equivalent from the image theory of electromagnetics.

The microstrip quasi-horn antenna and other quasi-horn antennas realized on unbalanced transmission lines do not need a transition or balun at the antenna input. Due to its microstrip nature, the top conductor of the microstrip quasi-horn antenna may also be fabricated together with other microstrip circuits on a single substrate by a photolithography process. In this integration, a common ground plane can be used and the antenna portion would be raised above the ground plane and supported by another dielectric (e.g., Styrofoam). The antenna is thus suitable for low-cost production. When two microstrip quasi-horn antennas are placed against each other, the common ground plane can act as a shield between these antennas, resulting in (theoretically) inherent isolation between them. This unique feature is extremely attractive for system applications where a certain degree of isolation is always needed between transmit and receive antennas. For other antennas used in practice, an inherent coupling between transmit and receive antennas always exists. Spacing these antennas apart, placing an absorbing material or a conductor wall between them can reduce the coupling, but also increase the system size and cost, as well as encounter possible problems in the antenna alignment, packaging, and reproducibility.

The operation of the microstrip quasi-horn antenna is basically based on the principle of wave propagation along a transmission line. In the uniform section of the microstrip line, where the spacing between the top conductor and the ground plane is very small compared to a wavelength, the wave propagation is mostly confined within the dielectric between the top conductor and the ground plane. However, as the separation between the conductor and ground plane gradually increases and approaches a half-wavelength or more, the energy begins to radiate in end-fire mode and so the wave is no longer guided between the conductor and ground plane. The entire structure effectively behaves as an antenna. The width of the antenna aperture determines the radiation efficiency and sets the low-frequency limit, while the antenna length and the dielectric contour control the matching over an operating bandwidth. So by properly and gradually tapering the dielectric substrate, a microstrip line can serve as a very wideband antenna, delivering energy from the feed point toward the open end at all operating frequencies with minimum reflection not only at the antenna input, but also along the antenna body and at the antenna aperture, which is highly desirable for the transmission and reception of pulse signals in UWB systems, as mentioned earlier.

A rigorous design approach is applied for the microstrip quasi-horn antenna following the TEM horn antenna design described in [12]. The design method optimizes the antenna dimensions for a given input reflection and minimizes the pulse stretching effect on the antenna. We define the coordinates including the x- and r-axis along the top conductor as indicated in Fig. 5.1. It is recognized that the E-plane of the microstrip quasi-horn is along the θ-direction and the H-plane is along the ϕ-direction, according to the notation defined in Fig. 5.1.

The non-uniform microstrip line used in the microstrip quasi-horn antenna can be designed in the same way of a non-uniform or tapered transmission line used for impedance matching between two different impedances. The optimum characteristic impedance variation of a non-uniform transmission line for a given maximum allowable input reflection coefficient was derived in [13] as

$$\ln Z(x) = \frac{1}{2}\ln\left[Z(0)Z(X)\right] + \frac{1}{2}\ln\left[\frac{Z(X)}{Z(0)}\right]G\left(B,\frac{2x}{X}\right) \tag{5.1}$$

where $Z(x)$ represents the characteristic impedance at position x along the transmission line, and the function G is defined as

$$G(B,\xi) = \frac{B}{\sinh B}\int_0^\xi I_0\left(B\sqrt{1-\xi'^2}\right)d\xi' \tag{5.2}$$

where $\xi = 2x/X$, X is the length of the non-uniform transmission line, which is the length of the antenna in the x-direction as defined here, I_0 is the Bessel function of the first kind of zero order, and B is the parameter relating to the maximum allowable magnitude of input reflection coefficient R(0) as

$$|R(0)|_{max} = \tanh\left[\frac{B}{\sinh B}(0.21723)\sqrt{\left\{\frac{Z(X)}{Z(0)}\right\}}\right] \tag{5.3}$$

The parameter B is also related to the length and the lower frequency limit of the antenna such that

$$\beta_{min} X = \sqrt{(B^2 + 6.523)} \tag{5.4}$$

where β_{min} represents the phase constant at the lower frequency limit f_{min}. For the special of $B=0$, the shape of the non-uniform transmission line becomes the commonly used exponential taper.

As indicated in [12], pulse signals radiated through a TEM horn is stretched, owing to different traveling paths of the signals along the antenna structure, and the pulse stretching is minimized in the boresight direction ($\theta=0$, $\phi=0$). A simple equation for the maximum pulse stretching time in the boresight direction on the E-plane was derived in [14] as

$$t_{max} = \frac{L}{u}(1 - \cos a) \tag{5.5}$$

where u is the phase velocity of the propagating TEM wave and a is the apex angle between the x- and r-axis. This result implies that the maximum pulse stretching comes from the two most different traveling paths along the antenna: one is the path along the conductor and another is the path along the center axis of the antenna. Eq. (5.5) can also be used for the microstrip quasi-horn antenna design to obtain approximate result.

The design of the microstrip quasi-horn antenna can be performed in the following three steps:

Step 1: Determination of Antenna Length and Characteristic Impedances In the first step, the length of the antenna and the variation of the characteristic impedance of the microstrip line along the antenna length are determined based on Eqs. (5.1–5.5). Assuming that the maximum allowable input reflection is $|R(0)|_{max} = 0.3$, which corresponds to 10 dB of input return loss. For this input reflection, the parameter B can be determined as 2.1 using (5.3). Assuming further that the lower limiting frequency of the antenna is 0.3 GHz and the apex angle $a = 10°$, the required length of the antenna, X, can be calculated as 0.526 m using (5.4). The total length of the antenna in the radial r-direction, L, is calculated as 0.534 m from $X = Lcos(a)$. Using (5.5), the calculated pulse stretching time in the boresight direction is only 27 ps which is an acceptable value for our design; this value also implies that the assumption of 10° for the apex angle is reasonable.

To calculate the characteristic impedance variance along the top conductor, $Z(x)$, using (5.1), we need to know not only the required values of the parameter B and antenna length X, but also the value of the terminating characteristic impedance $Z(X)$ at the open end of the antenna, that enables optimum matching, and hence optimum wave radiation at the antenna aperture. The free-space intrinsic impedance is 377 ohms. This impedance, however, may not be the optimum value for the terminating impedance $Z(X)$. In this design, the optimum value of $Z(X)$ is determined through simulations of various antennas designed for different values of $Z(X)$ using the EM simulator Microwave Studio [15]. The values for $Z(x)$ are calculated using (5.1) and (5.2).

Step 2: Determination of Antenna Shape The second step in the antenna design is to determine the shape of the antenna (or the shape of the dielectric region) along the x-direction with respect to the h-direction, which in turn determines the distance h(x) between the conductor and the ground plane along the antenna length. It is important to avoid any abrupt transition in the shape along the entire antenna length to minimize undesirable reflections along the antenna length. The choice of the shape function is essentially the same as that used in impedance matching using non-uniform or tapered transmission line. Different shape functions would produce different transitions between adjacent points along the length. Nevertheless, virtually any shape with no-abrupt transition would produce similar matching if sufficient length is used for the antenna. Several different shapes for the microstrip quasi-horn antenna were investigated in [16] and the results do not show any noticeable difference in matching for different designs. However, possible improvement in both impedance transformation and antenna radiation characteristic can be achieved by combining different antenna shapes (or, equivalently, distance functions) on the same antenna. A combination of the exponential and cosine-squared functions was used for a millimeter-wave microstrip quasi-horn antenna in [17] to improve the antenna's matching and radiation. In this antenna, the exponential function is used to determine the distance between the top conductor and the ground plane up to one-half wavelength of the lowest frequency, and the cosine-squared function is used for the distance from one-half wavelength to the open end. The characteristic-impedance change from the input port to the open end of the antenna is subject to an exponential taper. It is noted that maintaining the top conductor in parallel with

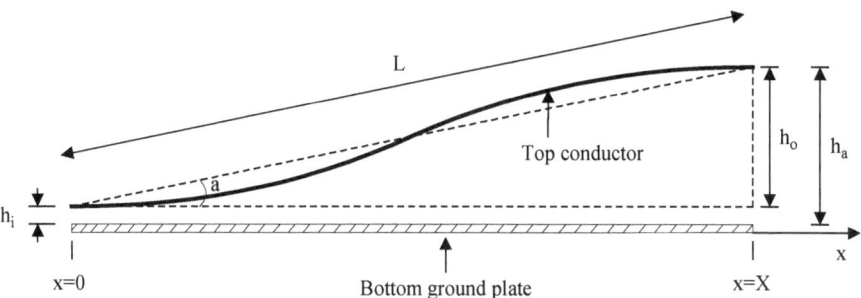

Fig. 5.2 Side view of the microstrip quasi-horn antenna with a sine-squared antenna shape

the ground plane at the open end of the antenna helps minimize the distortion of the radiating wave.

In our design, the antenna shape along the x-direction with respect to the h-direction, and hence the distance between the conductor and the ground plane, is selected as a sine-squared function as

$$h(x) = h_i + h_o \sin^2\left(\frac{\pi}{2}\frac{x}{X}\right) \qquad (5.6)$$

where h_i is the height at the input of the antenna, and $h_o = X \tan(a)$ and $h_a = h_i + h_o$ are the heights at the input of the antenna, as defined in Fig. 5.2. The resulting shape is similar to that shown in Fig. 5.2, representing a side view of the microstrip quasi-horn antenna. All the parameter values required for determining h(x) from (5.6) are already obtained earlier, except h_i. h_i is also important and related to possible higher-order-mode generation inside the antenna structure. For the conventional TEM horn antenna using the parallel-plate transmission line, there is a cutoff frequency pertaining to each higher-order mode. At frequencies higher than the cutoff frequency of a higher-order mode, the antenna may generate a corresponding higher-order-mode signal. The cutoff frequency is related to the plate separation between the top and bottom conductors at the input feeding of the antenna as

$$\lambda_c = \frac{2S}{n} \qquad (5.7)$$

where λ_c is the wavelength at the cutoff frequency, S is the plate separation at the input feeding, and n is the mode number of the higher order mode. In our case, the frequency range of the transmitting pulse is from 0.15 to 3.7 GHz. Taking into account some margin, the required cutoff frequency in our design is considered as 5 GHz. The mode number n is set to 1, assuming no higher-order modes larger than 1. The required separation at the feeding is then calculated as 30 mm using (5.7). In case of the microstrip quasi-horn antenna, the separation at feeding h_i is half of S; hence h_i should be less than 15 mm. In actual design, $h_i = 2$ mm is used to make a simple transition from the antenna input feeding to the SMA connector.

Table 5.1 Dimensions of the designed microstrip quasi-horn antenna

x-axis (mm)	Z(x) (Ω)	h(x) (mm)	W(x) (mm)
0.0	50.0	2.0	9.7
26.3	51.4	2.6	12.0
52.6	53.1	4.3	19.2
78.9	55.2	7.1	30.1
105.2	57.7	10.9	43.4
131.5	60.5	15.6	58.0
157.8	63.8	21.1	72.5
184.1	67.4	27.3	86.2
210.4	71.4	34.0	98.1
236.7	75.9	41.1	107.5
263.0	80.6	48.4	114.6
289.3	85.7	55.6	118.6
315.6	91.0	62.7	120.3
341.9	96.4	69.4	119.9
368.2	101.9	75.6	117.7
394.5	107.4	81.2	114.1
420.8	112.7	85.9	109.5
447.1	117.7	89.7	104.5
473.4	122.3	92.5	99.2
499.7	126.4	94.2	93.9
526.0	130.0	94.8	88.7

Step 3: Determination of Antenna's Top-Conductor Width The final step involves the determination of the width of the top conductor along the antenna length. This width is determined based on microstrip line, whose dielectric thickness is given by (5.6) and relative dielectric constant is according to the selected substrate, and the characteristic impedance Z(x) obtained in the first step. The number of points along the antenna length at which the width of the top conductor (or the characteristic impedance) is determined should be sufficiently large to ensure smooth transition for the antenna shape. To that end, 21 equal-spaced points are used. Table 5.1 shows the characteristic impedance Z(x), the height h(x), and the width W(x) at different points on the x-axis along the antenna length for the microstrip quasi-horn antenna designed for $|R(0)|_{max}$=0.3, B=2.1, f_{min}=0.3 GHz, Z(X)=130 Ω, and air as the dielectric. The antenna's input and output heights are h_i=2 mm and h_a=96.8 mm, respectively, and the antenna length is X=526 mm.

As indicated in Step 1 of the antenna design, the optimum terminating characteristic impedance Z(X) at the open end of the antenna is determined through EM simulations using Microwave Studio for different antennas designed for Z(X) values of 100, 130, 180, and 377 Ω. The results show that 130 and 377 Ω provide the best and worst performance for the antenna, respectively. Figure 5.3 shows two designed antenna structures selected to show these two extreme cases of design. The antenna in Fig. 5.3a is designed for Z(X)=377 Ω, which corresponds to the design with inferior performance, and the one in Fig 5.3b is designed for Z(X)=130 Ω which produces superior performance.

Fig. 5.3 Physical profile of
the microstrip quasi-horn
antenna for Z(X) of 377 Ω
a, and 130 Ω **b**

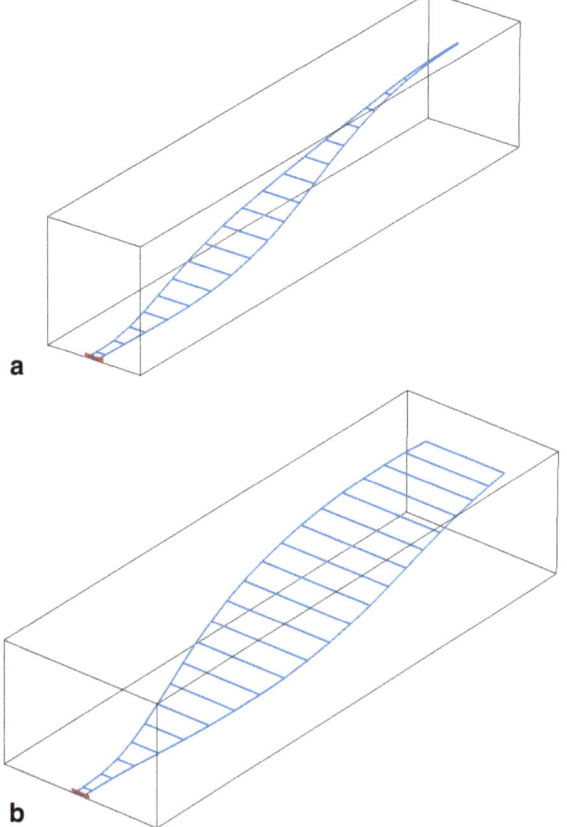

a

b

Figure 5.4 shows the input return losses obtained from the EM simulations for
the two designed antenna shown in Fig. 5.3. The results shown in Fig. 5.4 indicate
that the antenna structure of Fig. 5.3b with Z(X)=130 Ω has significantly better
performance for the input matching.

Figure 5.5 shows the EM-simulated time-domain reflections of a 400-ps monocycle pulse, seen at the input of the antenna, as it traverses the antenna. The initial
reflection at around 1 ns is at the antenna's input, while the last reflection at around
4.5 ns is at the open end of the antenna, and those in between are along the antenna
length between the input and output of the antenna. The results indicate that a significant reduction in the reflection at the antenna aperture is achieved for the design
with Z(X)=130 Ω. It is particularly noted that, while the frequency-domain return
loss as displayed in Fig. 5.4 shows the return loss at discrete frequencies (or return
loss of individual frequency components of the pulse), the time-domain return loss
obtained through the reflections shows the return loss of the entire pulse itself and
hence reflects truly the matching of the antenna used for UWB systems operated
based on pulse signals. This time-domain return losses, as calculated from Fig. 5.5,
is around 23 and 8 dB, and 21 and 16 dB at the antenna's feed point and aperture for
Z(X) of 377 and 130 Ω, respectively.

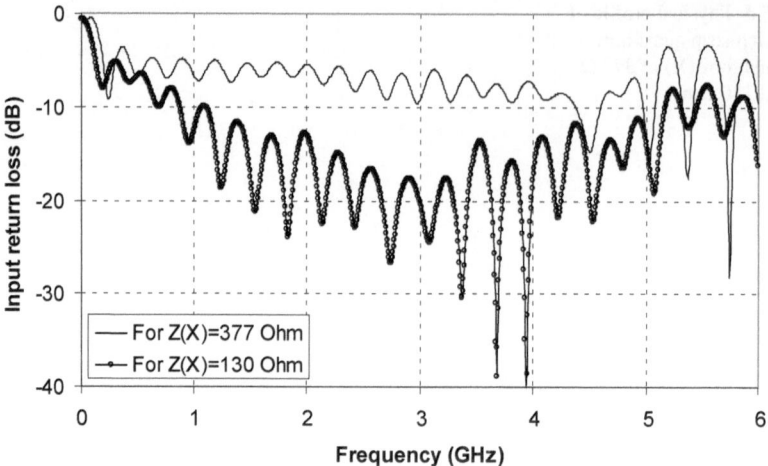

Fig. 5.4 Simulated input return losses of the designed microstrip quasi-horn antennas

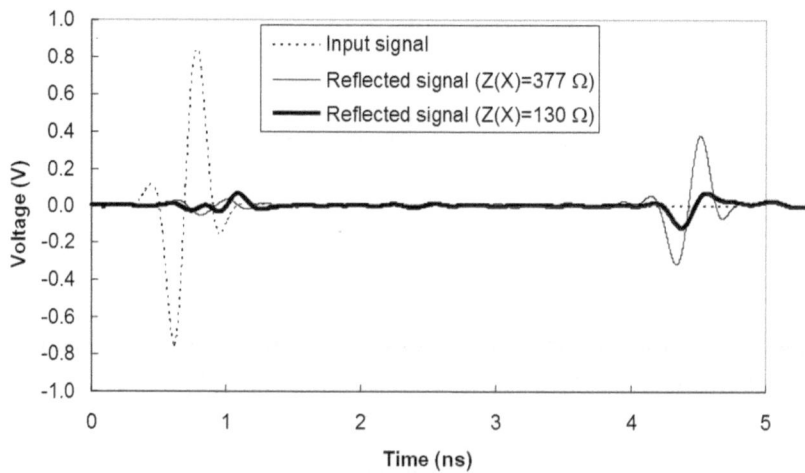

Fig. 5.5 Simulation results for the reflections of the microstrip quasi-horn antennas for an incident 400-ps monocycle pulse signal

It is noted again that, for time-domain applications, such as UWB systems which transmit and receive pulse signals, characterizing the antenna's transient responses directly is useful and important for accurate assessment of its signal fidelity and matching. Although from the Fourier series point of view, the frequency and time domain are correlated, and one can then view them as equivalent, they should be distinguished from one to another for UWB time-domain applications.

Figure 5.6 shows the calculated E- and H-plane radiation patterns at 1 and 2.5 GHz for the antenna designed with $Z(X) = 130 \, \Omega$. These results indicate that the

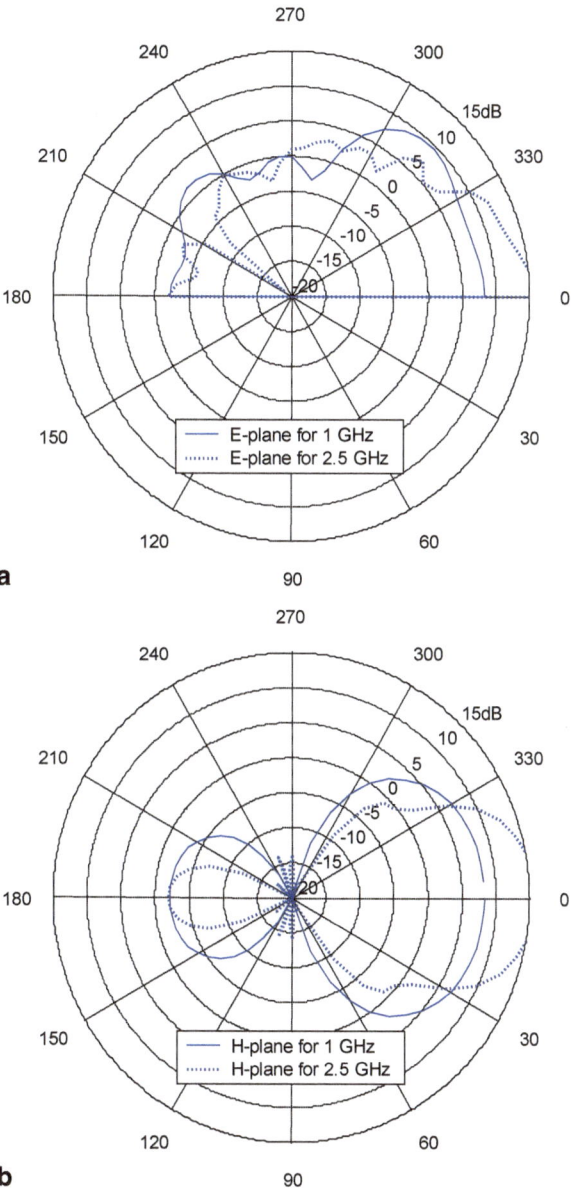

Fig. 5.6 Antenna pattern simulation results for the designed antenna with $Z(X) = 130\ \Omega$ shown in Fig. 5.3(b): E-plane patterns **a** and H-plane patterns **b** at 1 and 2.5 GHz

maximum antenna gain is more than 10 dB at 2.5 GHz, which is about the center frequency of the 450-ps monocycle pulse signal used in the UWB system presented in this book.

Figure 5.7 shows the calculated electric field intensity on the antenna at 10-cm away from the antenna input in the boresight direction for both $Z(X) = 130\ 377\ \Omega$ for a 400-ps monocycle pulse input signal. The waveform of the electric field is

Fig. 5.7 Simulation results
of the field intensity for the
designed microstrip quasi-
horn antennas with a 400-ps
monocycle pulse input signal

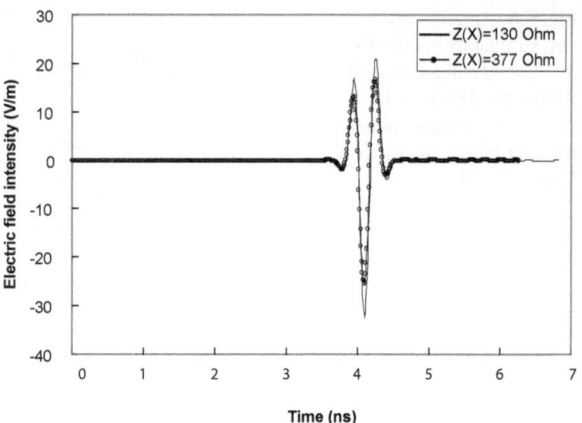

Fig. 5.8 Fabricated
microstrip quasi-horn antenna

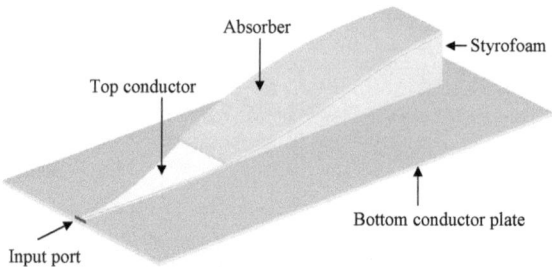

seen as the first derivative of the input monocycle pulse signal as it propagates
along the microstrip quasi-horn antenna, as mentioned in Sect. 2.5. The simulation
results indicate that the antenna designed with 130 Ω radiates pulse signal with
higher power and good waveform shape having relatively low side-lobe ringing.

From the foregoing simulation results, it is concluded that the optimum design
of the microstrip quasi horn antenna is the design shown in Table 5.1 and Fig. 5.3b
with the terminating impedance $Z(X) = 130\ \Omega$.

5.2.2 Fabrication and Performance of the Microstrip Quasi-Horn Antenna

Figure 5.8 shows the drawing of the fabricated microstrip quasi horn antenna designed using 130 Ω terminating impedance. The dielectric substrate is Styrofoam
having a relative dielectric constant around 1. It is basically the same structure as
the one in Fig. 5.3b, except that the fabricated antenna has absorbing material covering the top conductor, which helps improve the performance of the input matching
and antenna gain. The absorber used here is the laminated radar absorber, AEL-
0.375 manufactured by Advanced Electromagnetics.

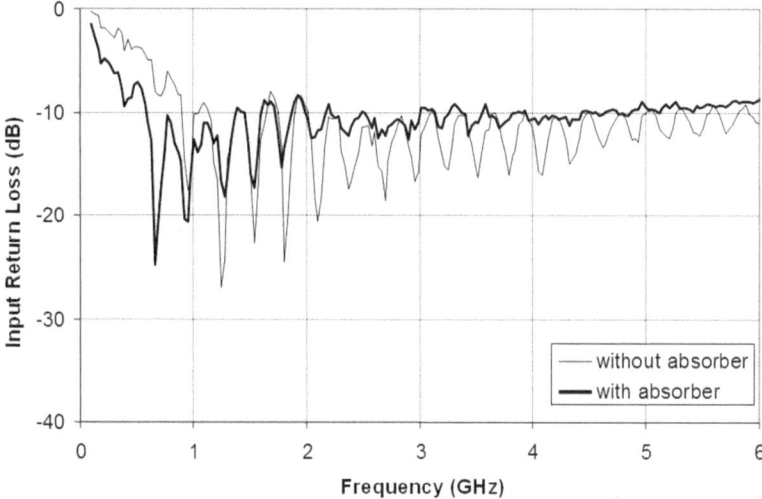

Fig. 5.9 Measured input return loss for the fabricated microstrip quasi-horn antenna

Figure 5.9 shows the measured input return loss of the designed antenna with and without absorber covering the top conductor. The measurement results show noticeable performance improvement with absorber at frequencies below 1 GHz. The results also indicate that the designed antenna with absorber supports at least 2:1 input VSWR for CW signals over the frequency range from 0.5 to 6 GHz.

As mentioned earlier, for UWB systems, the performance of the microstrip qua-si-horn antenna in the time domain is needed to assess its performance in radiating and receiving pulses, and is much more critical than its frequency counterpart for the design and evaluation of not only the antenna itself, but also of associated components and UWB systems.

Figure 5.10 shows the system block diagram used to measure the time-domain waveforms of the reflected pulse signals on the microstrip quasi-horn antenna for an input pulse signal. The transmitter in Fig. 5.10 is the designed UWB transmitter described in Chap. 3 which generates a monocycle pulse of varying pulse durations. This measurement setup enables us to measure the reflected signals from an input wideband pulse signal covering up to 12 GHz, which is the upper limiting frequency of the employed digitizing oscilloscope. An extra calibration process is needed to compensate for the loss due to the resistive power divider. As can be seen in the measurement setup shown in Fig. 5.10, a reflected signal traveling toward the oscilloscope would undergo 6 dB of additional power loss through the resistive power divider as compared to the transmitting signal at the antenna (input signal). The additional loss can be calibrated by separating the reflected signal portion from the recorded overall signal in the oscilloscope and then multiplying it by 2.

Figure 5.11 shows the measured waveform of the input monocycle pulse to the designed antenna and the signal reflected from the antenna for a 450-ps incident

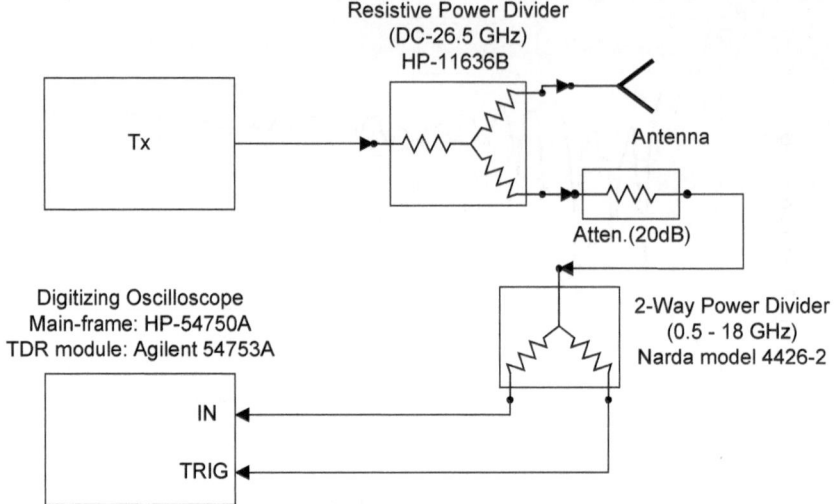

Fig. 5.10 Measurement setup for time-domain measurement of reflected signals in the microstrip quasi-horn antenna

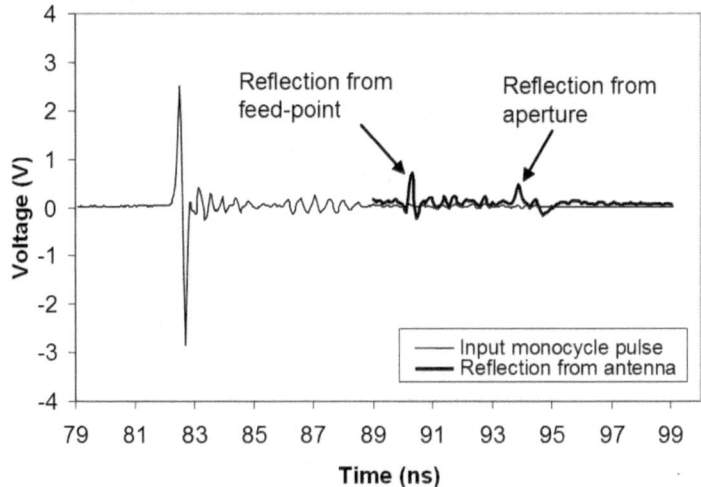

Fig. 5.11 Time-domain reflection measurement results

monocycle pulse signal. The reflected signals show two main reflections: one from the input port and another from the open end or aperture of the antenna. It is also observed that the voltage amplitudes of the reflections are comparable with the side-lobe ringing of the incident pulse, implying an acceptable performance for the input matching. The time-domain return losses at the antenna's feed point and aperture, as calculated from the reflected and incident pulses, are about 13 and 17 dB, respectively. It is noted that the input (time-domain) return loss is the actual return

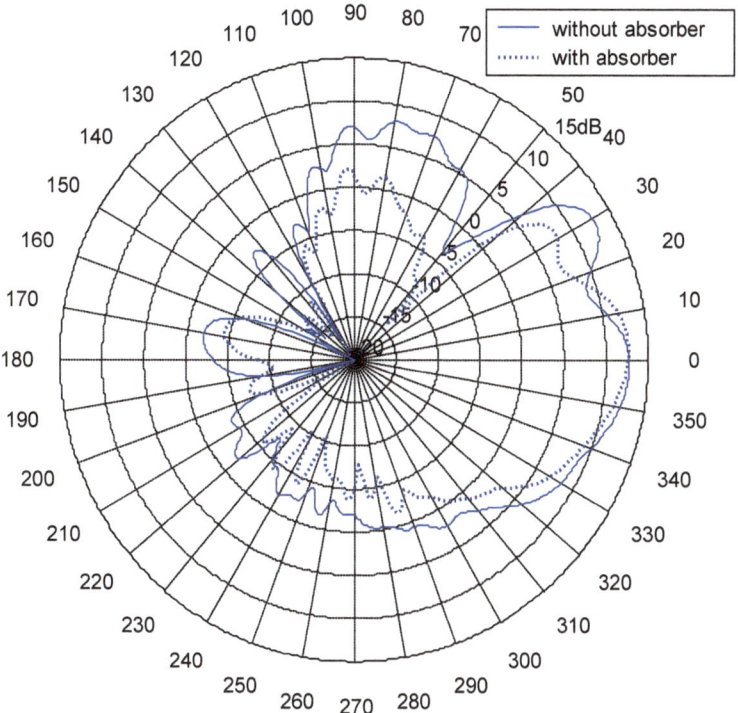

Fig. 5.12 Measured E-plane radiation patterns at 2.65 GHz

loss of the antenna used for time-domain UWB applications since it measures truly the antenna matching with respect to the entire input pulse signal. The return loss at the feed point is not as good as that at the aperture because it was difficult to maintain a correct spacing between the conductor and ground plane at the feed point corresponding to 50-ohm characteristic impedance, and this caused a small mismatch. This can be corrected easily by printing the conductor on a substrate and place the substrate flat against the ground plane at the feed point. The result also demonstrates that the internal reflections along the antenna structure (between the feed-point and aperture) are quite small. The negligibly transient internal reflections are obtained due to the smooth changes of the conductor width and Styrofoam height from the feed point to the aperture. Maintaining very small internal reflections is important for antennas operating in the time domain.

Figures 5.12, 5.13, and 5.14 display the measured radiation patterns of the antenna with and without absorber conducted in an antenna anechoic chamber. The measured antenna gain at 2.65 GHz is about 12 dBi with and without the absorber, while that at 4 GHz is around 15 and 12 dBi with and without the absorber, respectively. It can be seen that the use of absorber on the antenna results in increased antenna gain, especially at high frequencies. Although the radiation patterns were only measured at 2.65 GHz and 4 GHz, these results should be enough to show typical performance of the designed antenna. Moreover, the performance trend indicates

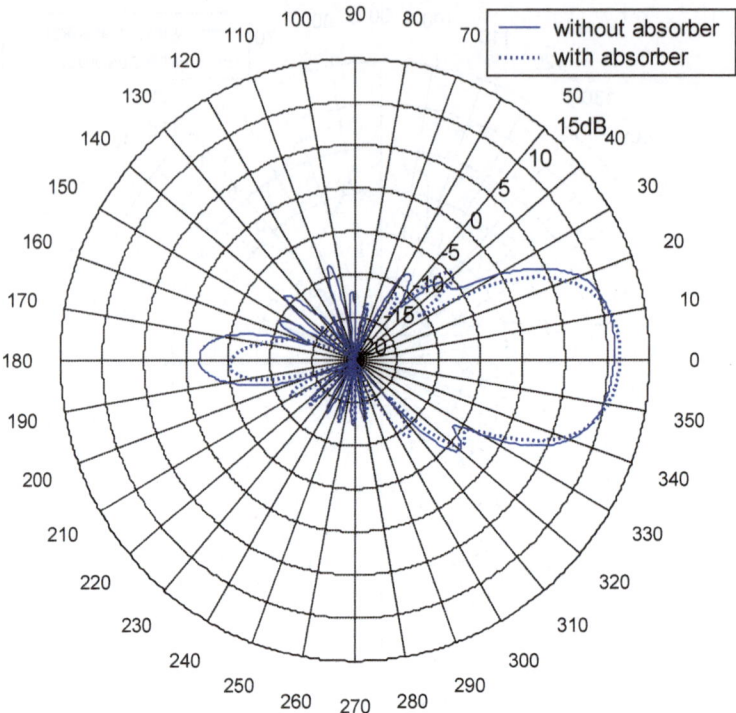

Fig. 5.13 Measured H-plane radiation patterns at 2.65 GHz

higher operating frequencies, which is also expected theoretically due to the increased size of the antenna with respect to the wavelengths at these frequencies. It should be noted that a large gain variation is possible at low frequencies than at high frequencies. This is expected because the antenna size is relatively small at the low frequency end, thus resulting in low gains. Using a larger size for the microstrip quasi-horn antenna would increase the gains at low frequencies and hence produce a small variation of the gain over the entire frequency range, but at an expense of increased size. It is observed that, whereas the maximum radiation in the H-plane is along the ground-plane axis, the maximum E-plane radiation tends to veer from that axis, which is also seen in [7]. The main beam axis is about 5° from the ground-plane axis in the E-plane. This is due to the asymmetrical structure of the antenna, resulting in an offset boresight angle.

The overall performance of the designed microstrip quasi-horn antenna was measured through a pulse transmission-reception test conducted indoor using various monocycle pulses of different durations. Figure 5.15 shows the measurement setup for the test. The transmitting and receiving antennas with identical structure face each other with a spacing of 4 ft 3 in. (20 cm). The designed UWB transmitter and the same resistive power divider and oscilloscope as shown in Fig. 5.10 are used here.

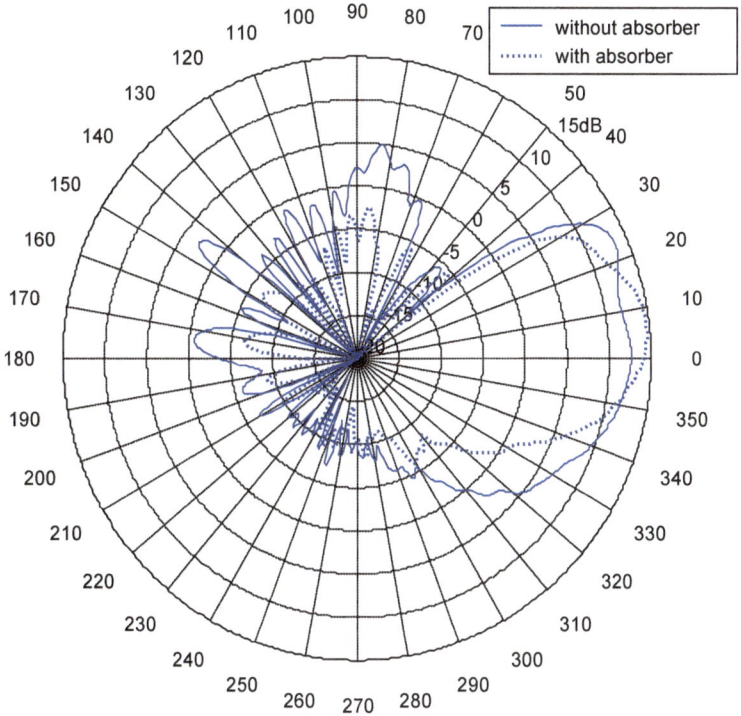

Fig. 5.14 Measured E-plane radiation patterns at 4 GHz

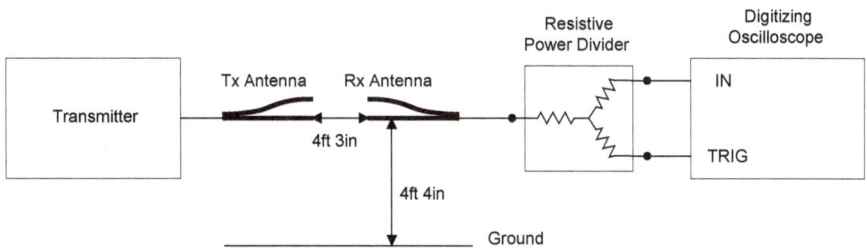

Fig. 5.15 System setup for the pulse transmission-reception test

Figure 5.16 shows the received pulse signals through the antennas displayed on the oscilloscope for four transmitting monocycle pulses with different pulse durations generated from the UWB transmitter. All the received waveforms follow the expected waveform shape that is the first derivative of the monocycle pulse as shown in Fig. 5.7, clearly demonstrating the antenna's ability to transmit and receive faithfully UWB impulse signals. There is no serious distortion in the received waveforms and almost no ringing except a small one coming from the original

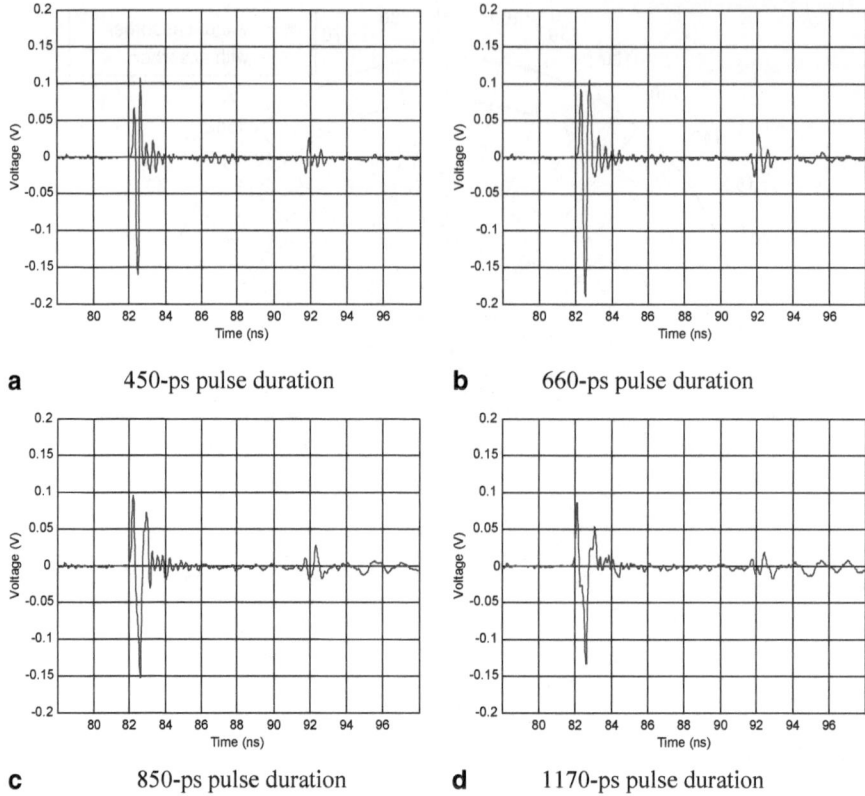

Fig. 5.16 Measured received signals in the pulse transmission-reception test for four transmitting monocycle pulses with different pulse durations. 450-ps pulse duration **a**, 660-ps pulse duration **b**, 850-ps pulse duration **c**, and 1170-ps pulse duration **d**

ringing at the tails of the transmitted pulses already discussed before. It should be noted again that the signal-fidelity characteristic of UWB antennas is critical for UWB systems. A spurious signal is also observed in each of the received waveforms around 10 ns after the main pulse, which is considered as a clutter signal from the surrounding structures in the indoor measurement environment. It is noticed that the maximum beam axis for the pulse signal was identified at approximately $\theta = 5°$ and $\phi = 0°$ through this experiment, which is coincident with the results of the antenna gain measurement.

The received pulse amplitudes as seen in Fig. 5.16 are considered reasonable through comparison with the calculated results of the received pulse amplitudes. The estimation of the received pulse amplitude is made possible by simply using the well-known signal-transmission equation:

$$P_r = P_t \frac{G_r G_t \lambda^2}{(4\pi R)^2} \tag{5.8}$$

Table 5.2 Measured performance of the microstrip quasi-horn antenna

Input VSWR (for CW signal measured in frequency domain)	2:1 (0.5 to 6 GHz)
Input return loss (for pulse signal measured in time-domain)	14 dB
Antenna gain (for CW signal)	12 dB @ 2.65 GHz
	15 dB @ 4 GHz
Antenna gain (for pulse signal)	10 dB
Main beam axis (for pulse signal)	$\theta = 5°$, $\phi = 0°$

where P_r is the available power at the receiver, P_t is the available power from the transmitter, λ is the operating wavelength, R is the distance between the transmitting and receiving antenna, and G_t and G_r represent the gains of the transmitting and receiving antennas, respectively. In our case, $G_t = G_r$. Although the measured antenna gain at 2.65 GHz for the designed microstrip quasi-horn antenna is about 12 dBi as seen in Figs. 5.12 and 5.14, the antenna gain is assumed to be 10 dBi in the calculations to take into account possible smaller gain for the entire wideband pulse signal. For the 450-ps pulse duration, the transmitter output pulse amplitude is 5.8 V_{p-p} (peak-to-peak voltage) which corresponds to 336 mW of RMS power level. Considering the center frequency as 2.65 GHz, the wavelength is $\lambda = 4.45$ in. Using (5.8) with R of 4-ft 3-in, the calculated available power at the receiver is 1.59 mW, which corresponds to 2 dBm. Assuming 1 dB of impedance-mismatching loss per antenna and also 1 dB of antenna radiation efficiency per antenna, the received power turns out to be -2 dBm, which corresponds to 251-mV_{p-p} amplitude of the received pulse. This calculation result is almost the same as the measurement result.

Table 5.2 summarizes the measured performance of the designed microstrip quasi-horn antenna.

5.3   UWB Uniplanar Antenna

5.3.1   UWB Uniplanar Antenna Design

Although the microstrip quasi-horn antenna presented in Sect. 5.2 is smaller than the conventional TEM horn antenna, its size is still relatively large for UWB systems designed completely on microwave integrated circuits (MICs), making these systems not convenient for portable or handheld uses. To reduce the size of UWB systems and facilitate integration with MIC transmitters and receivers, a compact UWB antenna with uniplanar structure, which can be fabricated completely on the same substrate with the transmitter and/or receiver, is preferred. In this section, we will present the design of such a low-cost, compact, easy-to-manufacture UWB uniplanar antenna for UWB systems.

Figure 5.17 shows the basic structure of the UWB uniplanar antenna which consists of two parallel non-uniform or tapered slot lines fed by CPW. This antenna can be considered as a planar version of the quasi-horn antenna which employs

Fig. 5.17 Basic structure of
the UWB uniplanar antenna

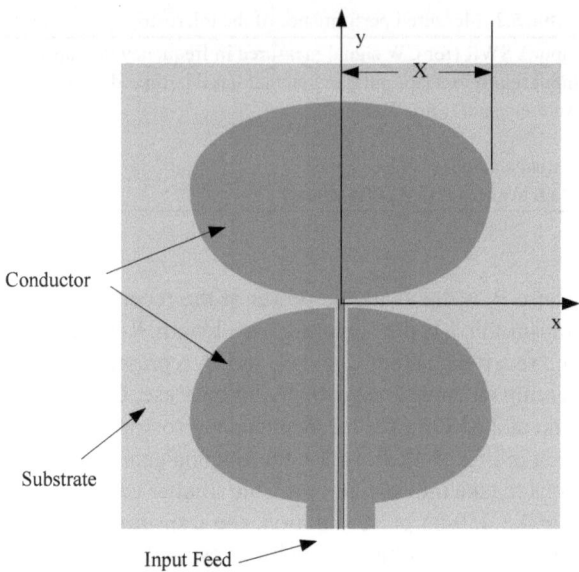

non-uniform slot line fabricated in two dimensions (2-D) instead of 3-D as for the
microstrip quasi-horn antenna. As such, the design of the UWB uniplanar antenna
can follow the design method for the microstrip quasi-horn antenna described in
Sect. 5.2. As for the microstrip quasi-horn antenna, the UWB uniplanar antenna
is fed from the side instead of at the center of its structure. A center-fed uniplanar
structure should be avoided due to the reason that the feed region of this structure
lies in the heart of the most intense near-fields surrounding the antenna. Strong
coupling between the feed structure and antenna seriously affects the near field and
distorts the antenna pattern [18].

In the uniform section of CPW lying inside one half of the antenna, most of the
energy is confined within the transmission line until it reaches the antenna center,
where the energy is coupled from the CPW to the two parallel non-uniform slot lines.
Assume the CPW has a typical characteristic impedance of 50 Ω, the slot line would
have a characteristic impedance of 100 Ω. The varying characteristic impedances
of the tapered slot lines, which produce minimum internal reflections along the slot
lines and input reflection not more than the maximum allowable input reflection for
the antenna, are calculated from Eq. (5.1) and the optimum terminating impedance
at the open end of the slot lines (and hence the antenna) can be determined via EM
simulations, as done for the microstrip quasi-horn antenna designed in the previous
section. The tapered slot lines essentially simulate an impedance transformer to
match 100 Ω to the open end's optimum terminating impedance. A remark needs
to be stated concerning the implementation of the tapered slot lines based on the
calculated varying characteristic impedances. A smooth-change contour, as shown
in Fig. 5.1, is essential to help avoid any abrupt transition in the shape across the
entire antenna structure to minimize undesirable reflections. For the non-uniform
slot-line transition sections close to the center of the antenna aperture, the gap width

is much smaller compared to the associated metal widths, which corresponds to small characteristic impedance; therefore the conventional analytical formula for the characteristic impedance of the slot line can be used to determine the gap width for the small characteristic impedance. On the contrary, for the non-uniform slot-line transition sections close to the open end of the antenna, the gap width keeps on increasing while the metal width reduces quickly, so the typical slot-line calculation method cannot be applied to the transition structure anymore; instead an EM simulator should be used to accurately determine the gap width for a specific metal width and required characteristic impedance. It should also be noted that, due to the abrupt variation of the slot edge around the open end of the antenna, the points on the tapered slot lines (along the x-axis on Fig. 5.17), where the characteristic impedances are calculated, should be chosen with non-uniform steps, which are sparse near the antenna aperture's center and condensed around the open end, to maintain a good variation for the characteristic impedances.

In our design, the antenna substrate is Duroid having a thickness of 0.635 cm and a relative dielectric constant of 10.5, and the impedance at the input feeding of the antenna is 50 Ω. The resulting characteristic impedance of the slot line at the antenna center is 100 Ω, which also facilitates the etching of the antenna pattern for the employed substrate parameters. If the characteristic impedance of the slot line is less than 100 Ω, the slot width would be too narrow and hence is difficult to be fabricated. On the other hand, the transition of the 50 Ω CPW to the two parallel 100 Ω slot lines should be as smooth as possible to minimize possible reflection due to physical discontinuity. To that end, the slot width of the slot line should be close to the gap width of the CPW.

Table 5.3 summarizes the slot-line's characteristic impedance $Z(x)$ along the x-axis and the corresponding dimensions of gap width $G(x)$ and metal width $W(x)$ (along the y-direction in Fig. 5.17) of the designed UWB uniplanar antenna. The dimensions were calculated using the EM simulator IE3D [19].

Microwave Studio [15] was used to perform the time-domain EM simulation and to optimize the antenna structure to minimize reflections occurring at the open-end transition. Figure 5.18 shows the simulated return loss of the antenna structure, as shown in Fig. 5.17, in the frequency domain. The result shows good return loss of more than 15 dB for the UWB uniplanar antenna over the frequency range of 3 to 12 GHz. The corresponding time-domain reflection result is presented in Fig. 5.19, where a Gaussian monocycle pulse with the 50 % pulse width of 50 ps is used as the excitation input signal. As shown in Fig. 5.19, only a small reflected signal occurs, hence validating the design of the antenna.

The transfer function of the UWB uniplanar antenna was also simulated at the position of 2 inches directly above the center of the antenna aperture surface. Figures 5.20 and 5.21 show the simulated results. The antenna transfer-function amplitude in Fig. 5.20 is normalized and has a band-pass response. The phase of the antenna transfer function, as shown in Fig. 5.21, indicates good linearity across 2–12 GHz, which should lead to little distortion to the waveform shape of the pulse signals transmitted and received by the antenna.

Table 5.3 Characteristic impedances and dimensions of the slot lines in the designed UWB uni-planar antenna

x (inch)	Z(x) (Ω)	G(x) (inch)	W(x) (inch)
0.010	100	0.030	0.750
0.0588	101	0.0324	0.7474
0.1171	103	0.0372	0.7406
0.1742	107	0.0528	0.7257
0.2296	113	0.0786	0.7029
0.2828	123	0.1082	0.6755
0.3333	139	0.1456	0.6416
0.3806	151	0.1902	0.6018
0.4243	165	0.242	0.5561
0.4638	178	0.313	0.4988
0.4989	193	0.392	435.7
0.5292	208	0.4778	0.3675
0.5543	226	0.5656	0.297
0.5742	240	0.6548	0.2247
0.5887	251	0.7484	0.1493
0.5971	257	0.8398	0.745
0.600	260	0.930	0

Fig. 5.18 Simulated return loss of the designed UWB uniplanar antenna

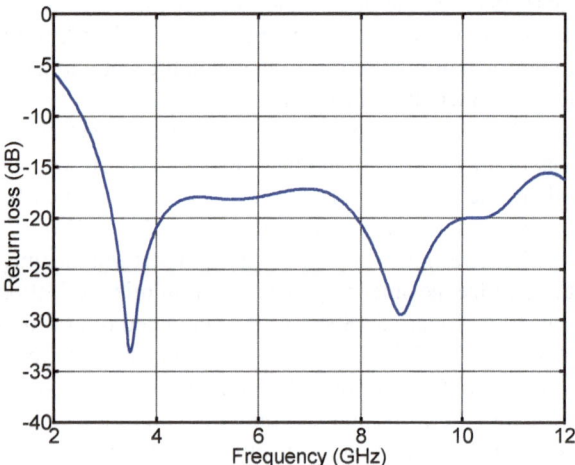

Fig. 5.19 Simulated input reflection of the designed UWB uniplanar antenna in time-domain

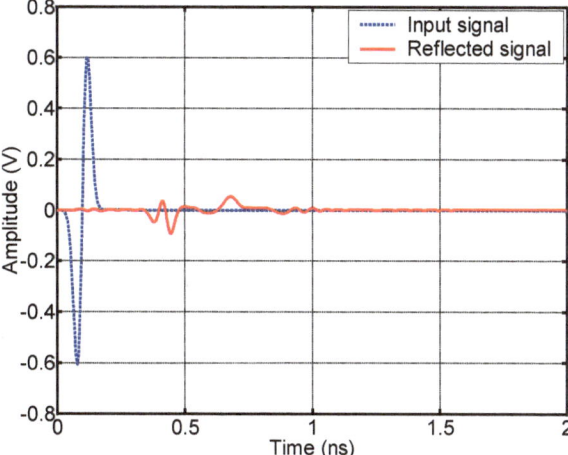

Fig. 5.20 Simulated amplitude of the transfer function of the designed UWB uniplanar antenna

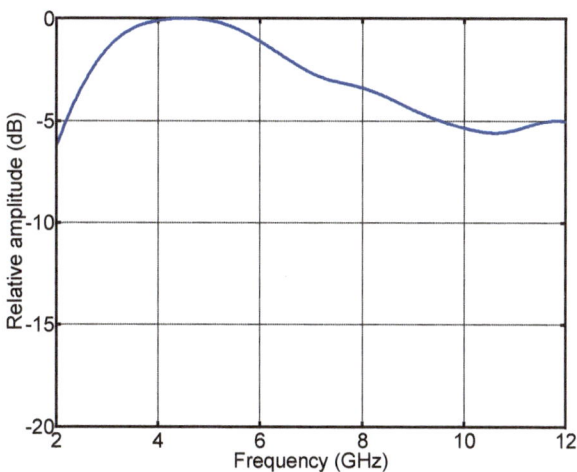

Fig. 5.21 Simulated phase of the transfer function of the designed UWB uniplanar antenna

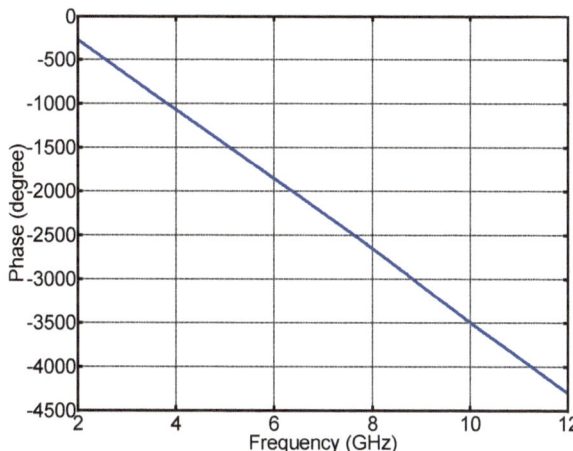

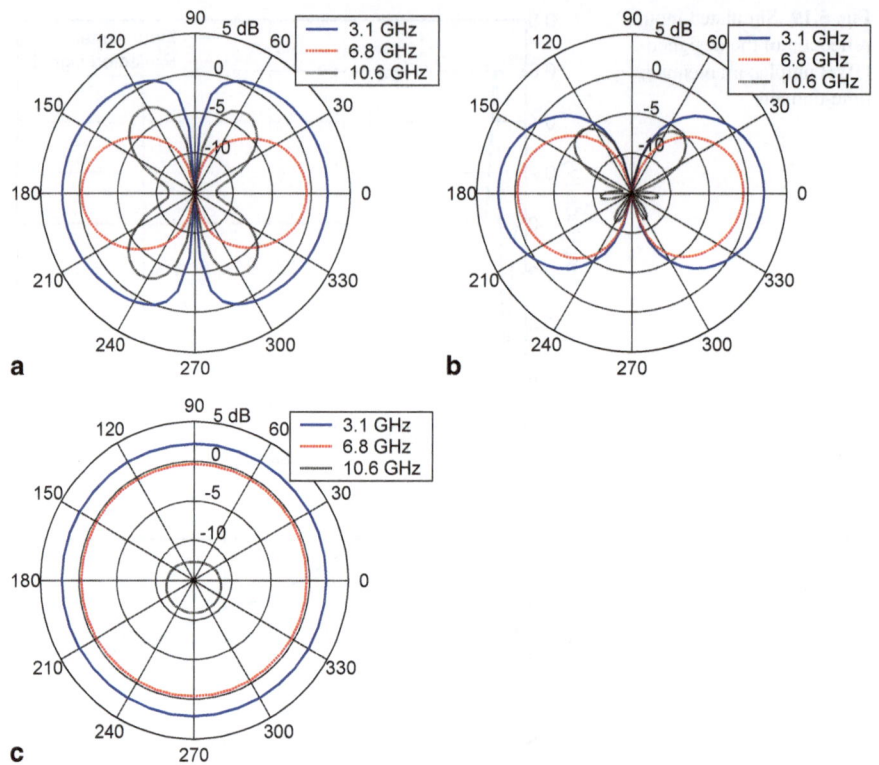

Fig. 5.22 Simulated antenna patterns for **a** *E*-plane (*x-y* plane), **b** *E*-plane (*y-z* plane), and **c** *H*-plane patterns of the designed UWB uniplanar antenna at 3.1, 6.8, and 10.6 GHz

Figure 5.22 shows the simulated radiation patterns for the of the designed UWB uniplanar antenna at 3.1, 6.8 and 10.6 GHz.

5.3.2   Fabrication and Performance of the UWB Uniplanar Antenna

Figure 5.23 shows a photograph of the fabricated UWB uniplanar antenna, including the antenna aperture, SMA fixture, and section of uniform 50 Ω CPW connecting the SMA connector to the antenna. The area occupied by the antenna aperture is only 1.2 in × 1.5 in.

Figure 5.24 shows the measured and simulated return losses in the frequency domain. Measured result shows more than 11-dB return loss from 3 to 12 GHz. As this return loss includes all the effects from the designed antenna, CPW feed line, and SMA connector, it is difficult to derive the antenna's actual performance from the frequency-domain results. On the contrary, it is relatively very easy to distinguish

Fig. 5.23 Photograph of the designed UWB uniplanar antenna along with the 50-Ω CPW feed line and SMA connector (on the left)

Fig. 5.24 Measured and simulated return loss of the UWB uniplanar antenna

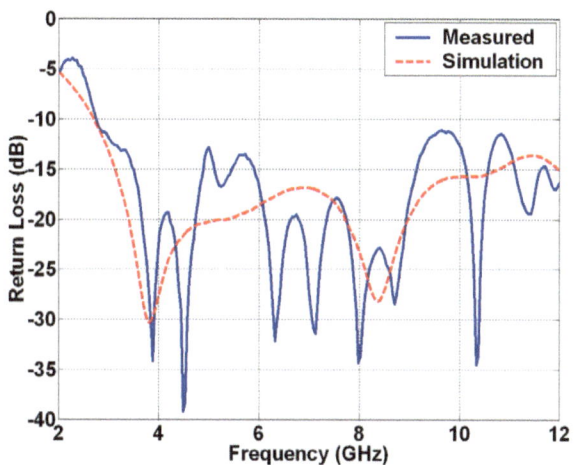

the antenna performance from other effects in the time domain. Furthermore, as the antenna is intended for radiating impulse or monocycle pulses for UWB systems, it is imperative to characterize it in the time domain. Figure 5.25 shows the measured and simulated time-domain reflectometry (TDR) response results in the time domain for a 50-ps input impulse signal. It is clear that, from 0 to 0.5 ns, the response corresponds to the effects of the SMA connector and CPW feed line. The response after 0.5 ns is caused by the designed antenna aperture and, as can be seen, the measured result matches very well with that simulated, which confirms the antenna design. The TDR performance also demonstrates excellent time-domain behavior of the designed antenna, which is crucial for time-domain UWB applications. The measured time-domain results indicate that better than 18-dB return loss is achieved for the antenna. This return loss is the true return loss for pulse signals and a better representation of the antenna matching than the frequency-domain return loss in Fig. 5.24. Good performance together with small size and uniplanar structure make the designed antenna a very good candidate not only for UWB applications but also for integration with printed-circuit UWB transmitters and receivers.

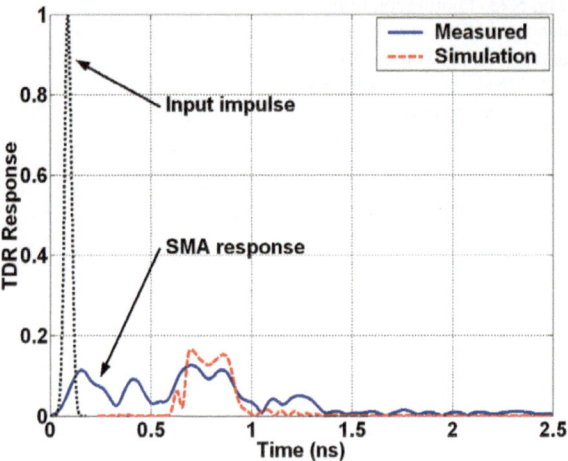

Fig. 5.25 Measured and calculated TDR responses of the UWB uniplanar antenna

To further evaluate and demonstrate the designed UWB uniplanar antenna for UWB applications, we conducted a pulse transmission test for the antenna. Figure 5.26 shows the block diagram of the test setup used for the pulse transmission measurement. A microstrip quasi-horn antenna operating from 0.2 to more than 20 GHz [7], designed based on the procedure discussed in Sect. 5.3, is used as the receiving antenna since it can produce faithfully the waveform of the received UWB signal. This signal fidelity of the microstrip quasi-horn antenna is demonstrated in [7] as well as in Sect. 5.2. The designed UWB uniplanar antenna and the microstrip quasi-horn antenna face each other and are spaced 3-ft apart. The pulse generator used in Fig. 5.26 was designed using CMOS radio-frequency integrated-circuit (RFIC) technology and can generate impulse or monocycle pulse signals. The impulse signals have 0.5–1.3 V peak-to-peak voltage with 100–300 ps tunable pulse duration (defined at 50 % of the peak amplitude). The monocycle pulses have 0.3–0.6 V peak-to-peak voltage and 140–350 ps tunable pulse duration at 50 % of the peak amplitude. The pulse received by the microstrip quasi-horn antenna is fed into a power divider and displayed in a 50-GHz digitizing oscilloscope.

Figure 5.27 shows the pulse signals received from the tunable impulse signals produced by the pulse generator and transmitted by the UWB uniplanar antenna. The pulse-duration tune-ability is clearly visible in the received pulses. As can be seen, the received signals are monocycle pulses with pulse duration tunable from 160 to 350 ps. The resultant monocycle waveform is due to the differential function of the designed antenna. The received pulses maintain good symmetry with no serious distortion and ringing.

Figure 5.28 shows the received pulse signal corresponding to the tunable monocycle pulse signals produced by the pulse generator and transmitted by the UWB uniplanar antenna. The received pulse also has varying durations. All the received signals have shape similar to the first derivative of the monocycle pulses, as expected from the designed antenna. Both the measured impulse and monocycle-

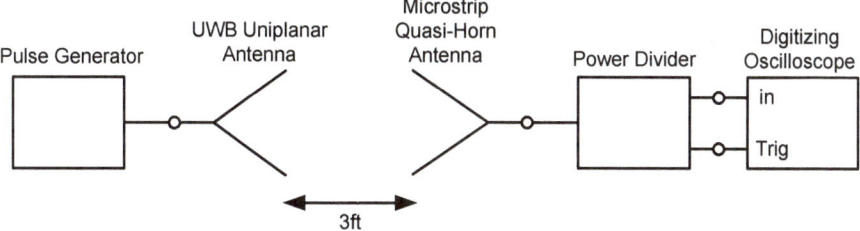

Fig. 5.26 Test setup for pulse transmission measurement

Fig. 5.27 Measured received signals of the impulses transmitted by the UWB uniplanar antenna

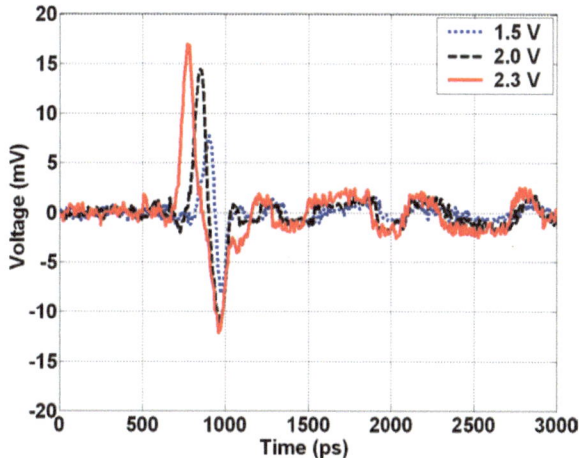

Fig. 5.28 Measured received signals of the monocycle pulses transmitted by the UWB uniplanar antenna

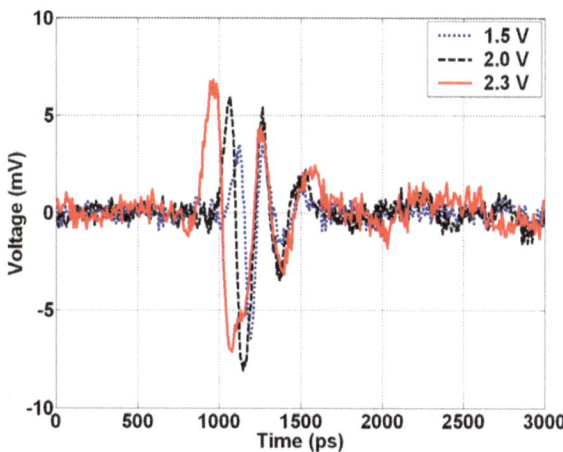

pulse transmission results clearly demonstrate the workability of the designed UWB uniplanar antenna for UWB applications.

5.4   Summary

The design of the quasi-horn antennas for UWB systems, which have extremely wide bandwidth, little distortion, and no balun or transition at the antenna input and can be directly integrated with printed-circuit UWB transmitters and receivers, has been presented. Specifically, two different UWB antennas were designed, fabricated, and tested: the UWB microstrip quasi-horn antenna and the UWB uniplanar antenna.

The microstrip quasi-horn antenna, designed for the transmission and reception of monocycle pulse signals with pulse duration in the range of 400 to 1200 ps, shows measured return loss of 13 dB for pulse operation and gain greater than 12 dBi at frequencies higher than 2.5 GHz. The antenna can transmit and receive 400 to 1200 ps monocycle pulses with small waveform distortion and ringing, demonstrating the antenna's good linear-phase response and low distortion, and hence is particularly desirable for time-main UWB applications.

The designed UWB uniplanar antenna can transmit and receive faithfully impulse signals having duration from 100 to 300 ps and monocycle pulse signals with duration of 140–350 ps. The antenna exhibits measured return loss of better than 18 dB for pulse operation. The experimental results for the impulse and monocycle-pulse transmission show the antenna's high signal fidelity through less-distorted received waveforms, demonstrating the workability of the antenna for UWB applications.

Reference

1. Theodorou, E.A., et al.: Broadband pulse-optimized antenna. IEEE Proc. Pt. H. **128**(3), 124–130 (June 1981)
2. Evans, S., Kong, F.N.: TEM horn antenna: Input reflection characteristics in transmission. IEEE Proc. Pt. H. **130**(6), 403–409 (October 1983)
3. Andrews, J.R.: UWB signal sources, antennas and propagation. Picosecond Pulse Labs, Boulder, CO, Application Note AN-14 (February 2003)
4. Schantz, H.G.: Introduction to ultra-wideband antennas. IEEE Conference on Ultra Wideband Systems and Technologies, 1–9 (16–19 November 2003)
5. Noronha, J.A.N., Bielawa, T., Anderson, C.R., Sweeney, D.G., Licul, S., Davis, W.A.: Designing antennas for UWB systems. Microwaves & RF (June 2003)
6. Cermignani, J.D., Madonna, R.G., Scheno, P.J., Anderson, J.: Measurement of the performance of a cavity backed exponentially flared TEM horn. Proc. SPIE: Ultrawideband Radar. **1631**, 146–154 (May 1992)
7. Nguyen, C., Lee, J.S., Park, J.S.: Novel ultra-wideband microstrip quasi-horn antenna. Electron. Lett. **37**(12), 731–732 (7 June 2001)
8. Han, J.W., Nguyen, C.: Investigation of time-domain response of microstrip quasi horn antennas for UWB applications. Electron. Lett. **43**(1), 9–10 (4 January 2007)

9. Lee, J.S., Nguyen, C., Scullion, T.: A novel compact, low-cost impulse ground penetrating radar for nondestructive evaluation of pavements. IEEE Trans. Instrum. Meas. **53**, 1502–1509 (December 2004)

10. Han, J.W. Nguyen, C.: Development of a tunable multi-band UWB radar sensor and its applications to subsurface sensing. IEEE Sens. J. **7**(1), 51–58 (January 2007)

11. Miao, M., Nguyen, C.: On the development of an integrated CMOS-based UWB tunable–pulse transmit module. IEEE Trans. Microw. Theory Tech. **54**(10), 3681–3687 (October 2006)

12. Theodorou, E.A., Gorman, M.R., Rigg, P.R., Kong, F.N.: Broadband pulse-optimised antenna. IEEE Proc. pt. H. **128**(3), 124–130 (June 1981)

13. Hecken, R.P.: A near-optimum matching section without discontinuities. IEEE Trans. Microw. Theory Tech. **20**(11), 734–739 (November 1972)

14. Daneshvar, K., Howard, L.: High current nanosecond pulse generator. Proc. IEEE Southeastcon '89, 572–576 (1989)

15. CST Microwave Studio, CST of America Inc., Wellesley Hills, MA (2005)

16. Lee, J.S.: Design of high-frequency pulse subsurface penetrating radar for pavement assessment, Ph. D. Dissertation. Texas A&M University (December 2000)

17. Park, J.S., Nguyen, C.: Low-Cost wideband millimeter-wave antennas with seamless connection to printed circuits. 2003 Asia Pacific Microwave Conference, Seoul (November 2003)

18. Schantz, H.G.: Bottom fed planar elliptical UWB antennas. IEEE Conference on Ultra Wideband Systems and Technologies, 219–223 (November 2003)

19. IE3D, Zeland Softwave Inc., Fremont, CA (2006)

Chapter 6
UWB System Integration and Test

6.1 Introduction

In this chapter, we present the integration and test of a UWB system, which consists of the UWB transmitter, receiver, and antennas described in Chaps. 3 to 5, and demonstrate its use as a UWB system for subsurface sensing. Figure 6.1 shows the block diagram of this UWB system that is similar to the block diagram shown in Fig. 4.2 in Chap. 4. The UWB system is completely fabricated using microwave integrated circuits (MICs) having the transmitter and receiver integrated directly with antennas on a single package. The system can vary the transmitting pulse duration, thus effectively simulating a multi-band UWB system or multiple UWB systems working together consecutively. This unique tuning multi-band/multi-pulse capability provides flexibility in the operation and application of the system and enhances its ability in detecting and classifying targets in different environments. For instance, narrow pulses have wide operating bandwidths and, thus, can be used to obtain fine resolution required for the detection of closely located objects. On the other hand, a wider pulse contains a larger fraction of the signal's energy in the low-frequency components, which have relatively low propagation losses in media, hence providing a longer detection range. Such an UWB system, whose transmitting pulse duration can be changed, especially by an electronic means, would therefore have both advantages of increased penetration (or range) and fine range resolution and is thus attractive. The multiband feature can also be exploited to determine a suitable trade-off between fine resolution and deep penetration. Moreover, the multiband operation of the UWB system can render more target information by combining those produced by individual operating frequency bands for certain targets. The tunable multi-band is particularly useful for practical lossy and dispersive multilayered structures, such as pavements or those containing buried mines, as the multiple bands and pulses can be utilized to possibly minimize the effects of loss and dispersion, resulting in better target extraction.

The integration and test of the UWB system are done in the following four separate stages:

In the first stage, the UWB transmitter-antenna and UWB receiver-antenna modules, constructed by integrating the designed UWB transmitter, described in

C. Nguyen, J. Han, *Time-Domain Ultra-Wideband Radar, Sensor and Components*,
SpringerBriefs in Electrical and Computer Engineering,
DOI 10.1007/978-1-4614-9578-9_6, © Springer International Publishing Switzerland 2014

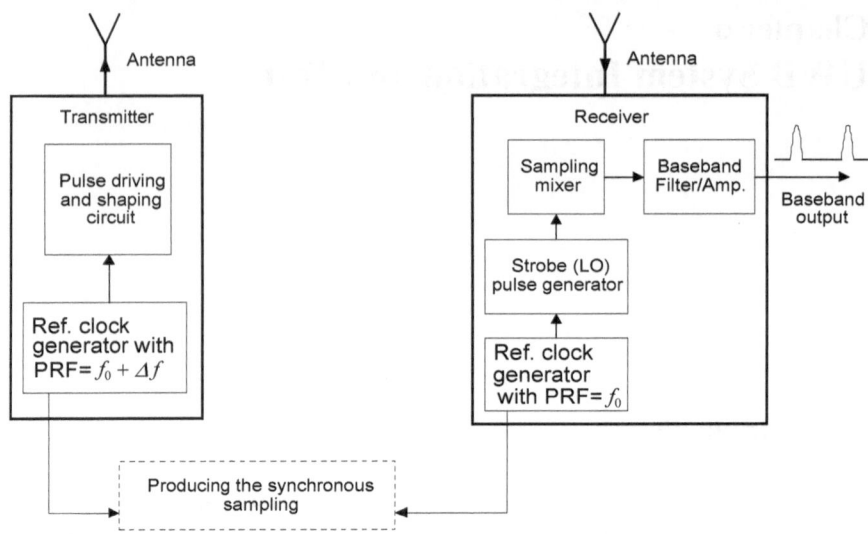

Fig. 6.1 Block diagram of the UWB system

Chap. 3, and UWB receiver, described in Chap. 4, with the UWB transmitting and receiving microstrip quasi-horn antennas, described in Chap. 5, respectively, were tested through a transmission-reception test to evaluate their performance. The transmission-reception test conducted here is similar to the transmission-reception test for antennas described in Chap. 5, but the designed UWB receiver was used to down-convert the received signal, which was displayed by a digitizing oscilloscope, instead of only using a digitizing oscilloscope for both down-conversion and display.

The second stage concerns the signal processing for the UWB system. The down-converted signal coming out from the UWB receiver is digitized by an analog-to-digital converter (ADC). The digitized signal normally contains both the cross-coupled signal from the transmitting antenna to the receiving antenna and the target reflected signal. A simple signal processing technique involving background subtraction is used to remove the cross-coupling signal in the received digital data. The background subtraction technique is a well-known technique in detecting target signals from input signals modulated by direct cross-coupling signals [1]. This technique is simple but very sensitive to jittering or frequency modulation (FM) effect in the detected baseband signal. The FM in the baseband signal is primarily due to the FM effect of the reference clock oscillators in the transmitter and receiver. Very stable clock oscillators are therefore required to reduce the FM effect in the baseband signal. A different approach will be introduced to overcome the FM problem without incurring the extra cost of expensive stable oscillators, which is a simple signal monitoring technique devised to improve the performance of signal processing based on background subtraction. The signal processing unit consists of the data

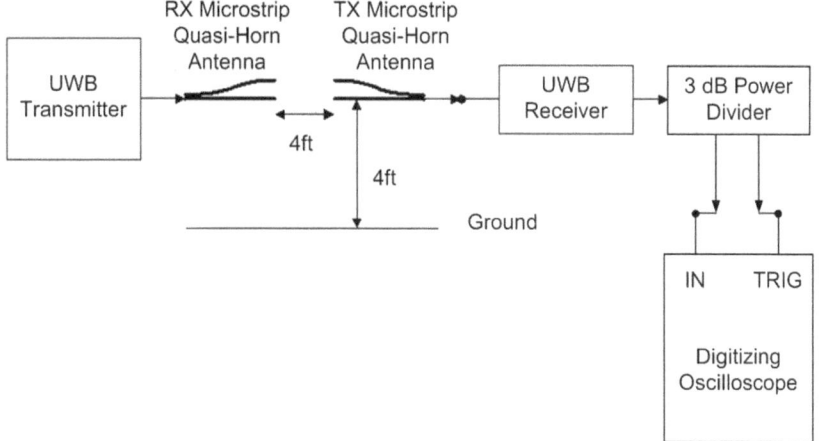

Fig. 6.2 Test setup for the transmission-reception test

acquisition card, DAQCard-6036E manufactured by National Instrument, and signal processing code written using Labview instrument programming language [2].

In the third stage, the complete UWB system was constructed by integrating the
UWB transmitter-antenna and UWB receiver-antenna modules into a single structure. The most important thing to be considered in this stage is the arrangement of
the two antennas, which affects the performance of the UWB system in detecting
signals reflected from targets. The best arrangement of the antennas supporting high
receiving power and low clutter signal level was obtained through experiments.

Finally, in the fourth stage, the performance of the UWB system, operated by
a notebook PC via the signal processing unit, was measured using several different target structures. These experimental results show the validity of the developed
UWB system directly, particularly as a UWB subsurface penetrating radar, and in-
directly the workability and usefulness of the designed UWB transmitter, receiver,
and antennas.

6.2   Transmission-Reception Test for UWB
Transmitter-Antenna and Receiver-Antenna Modules

Figure 6.2 shows the block diagram of the test setup used for the transmission-
reception test of the UWB transmitter-antenna and UWB receiver-antenna modules,
similar to that shown in Fig. 5.15 of Chap. 5 for antenna testing. The UWB transmitter-antenna and UWB receiver-antenna modules were placed 4 feet apart facing
each other and 4 feet from the floor in an indoor laboratory. It is noted that the 3-dB
power divider and the digitizing oscilloscope are used only for the measurement,
not as part of the designed UWB system. The transmitter was operated in the full-

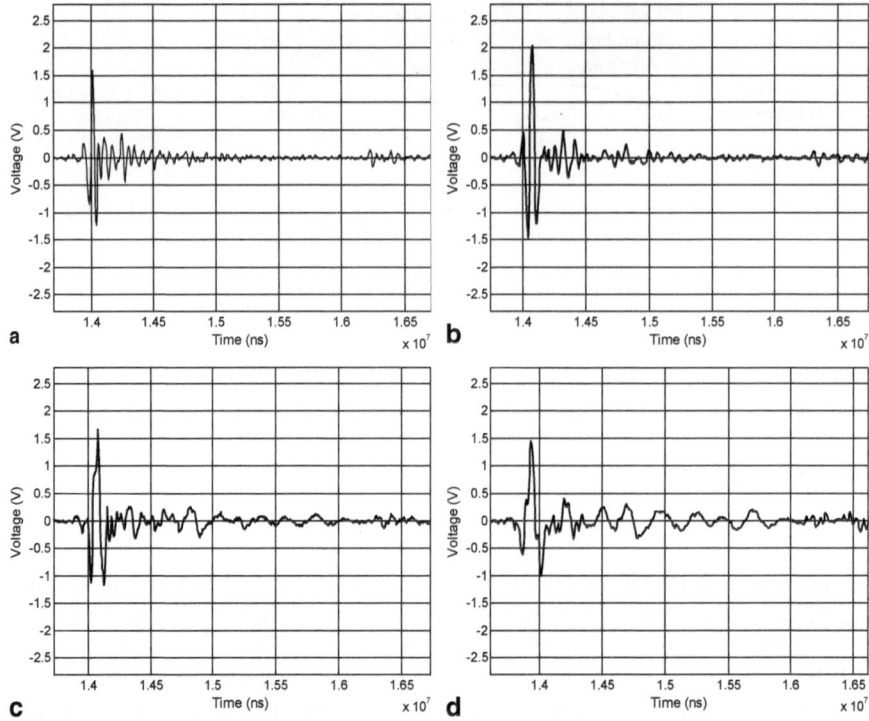

Fig. 6.3 Measured down-converted signals through the transmission-reception test for four different transmitting monocycle pulses of **a** 450 ps, **b** 600 ps, **c** 880 ps, and **d** 1170 ps

power mode. The main beam axes of both antennas were matched to each other for optimum transmission and reception.

The test was done using four different transmitting monocycle pulses having pulse durations of 450, 600, 880 and 1170 ps generated from the UWB transmitter. Figure 6.3 shows the waveforms of the down-converted pulse signals corresponding to these transmitting pulses that were recorded in the digitizing oscilloscope. As can be seen, the main lobes of the pulse waveforms do not show significant distortion, while there are some ringing in the mid- and late-time responses of the pulses, which effectively form the side-lobes of the pulses. These ringing were caused by some clutters around the UWB system in the indoor testing environment.

Notice that the unit of the time scale in Fig. 6.3 is 500 μs, which is due to the time scale transformation between the transmitted and received pulses. This time scale of the down-converted signal can be predicted as follows. For synchronous sampling, two reference clock oscillators are used in the transmitter and the receiver as discussed in Sects. 4.1 and 4.4 of Chap. 4. The deviation in frequency between the two clock oscillators is about 10 ppm, which corresponds to 100 Hz for a reference frequency of 10 MHz. As can be inferred from Eq. (4.7) in Chap. 4, a transmitting signal's single pulse repetition period of 100 ns is transformed into 10 ms in the time scale of its down-converted signal. Therefore, 1-ns pulse duration in the transmitting signal corresponds to 100-μs pulse duration in the down-converted signal. The time transformation can be seen more easily in Fig. 6.3c, where the 880-ps pulse duration occupies about a 100-μs time span.

6.3 Signal Processing

Signal processing is required to digitize the signal down-converted by the UWB receiver, to visualize the digitized data, and to detect the target signal from background noise or clutter signals. The signal processing unit was implemented using the National Instrument data acquisition (DAQ) card DAQCard-6036E, which has its own ADC, and Labview programming on a notebook PC. The Labview software package provides a convenient environment to build the signal processing and a variety of display formats. The data length of each digitized sample supported by the DAQ card is 16 bits, from which the dynamic range of the ADC can be calculated as $20\log(2^{16}) = 96$ dB. This dynamic range is much higher than the 50-dB dynamic range of the sampling mixer of the UWB receiver, and hence the quantization error induced by the ADC is negligible. The sampling rate for the input down-converted signal is set as 220 KHz according to the design parameter of the UWB synchronous sampling receiver derived in Chap. 4.

The digitized receiving data normally contain both the target reflected signal and the background cross-coupling signal, which is directly coupled from the transmitting antenna. Without removing this background cross-coupled signal (or background signal in short), it is hard to identify the true signal reflected from the target, which is amplitude modulated by the background signal. Herein, we employ a signal processing based on the background subtraction technique, which is a well-known method in impulse radar to detect desired signals. This technique, which is basically a demodulation process, consists of two simple steps: the first step of recording the background signal as a reference and then the second step of subtracting the reference from the real measured signal (or real signal) to extract the signal for the target structure. The use of the microstrip quasi-horn antennas for both transmitting and receiving antennas naturally enhances the isolation between them, as discussed in Sect. 5.1 of Chap. 5, thereby helping facilitate the background subtraction scheme. Assuming the reference signal is the same as the background signal included in the real signal, then there remains only the reflected signal from the target after the subtraction. The background subtraction method is simple but has the serious drawback of producing spurious detected signals if the background signals included in both the reference and the real signals are not identical. To resolve this problem, a method is introduced into our signal processing.

The background signal is first measured and recorded with the antenna pointing into the open space. This is essentially the calibration for the antenna to exclude any possible reflected signals apart from the direct cross-coupling signal between antennas. With this calibration, the background signal does not need to be measured

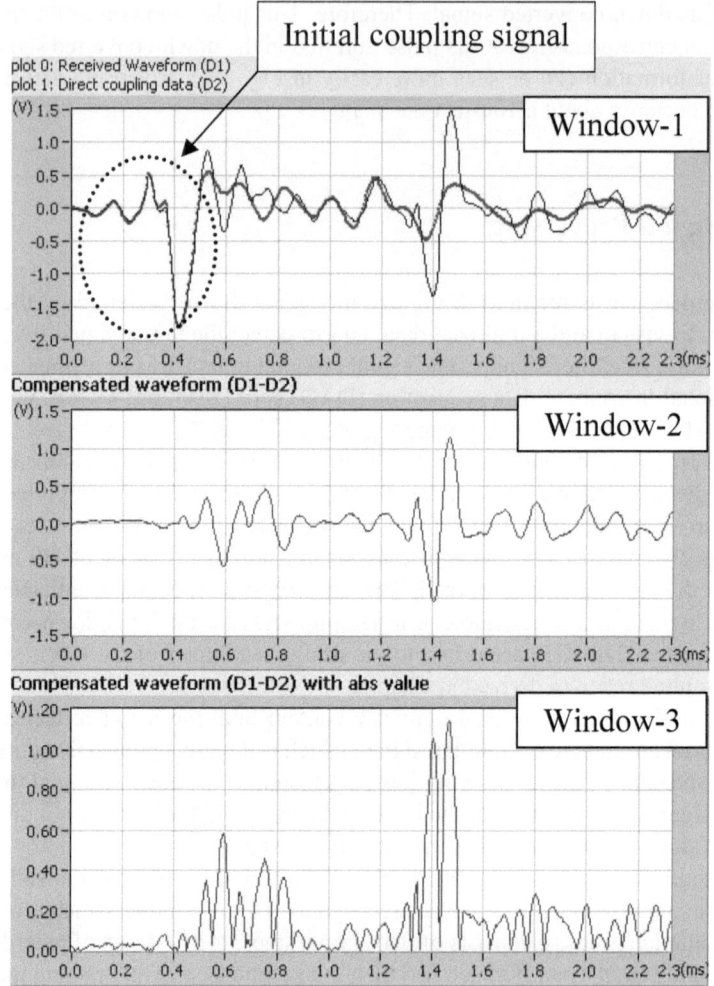

Fig. 6.4 Display window of the signal processing to detect a target's reflected signal using the background subtraction technique

every time real measurement is carried out; usually, a single accurate measurement of the background signal is enough.

Figure 6.4 shows a display window of the signal processing, programmed using Labview, to detect the reflected signal from a target by means of the background subtraction technique. The display window includes 3 sub-windows, each of which shows different data according to the processing steps.

Window-1 represents two signals: one is the reference signal represented by a thick line and another is the real signal represented by a thin line. The data shown in Window-1 play a very important role in the detection of the target's reflected signal. As seen in Window-1, the initial parts of the reference and the real signals' wave-

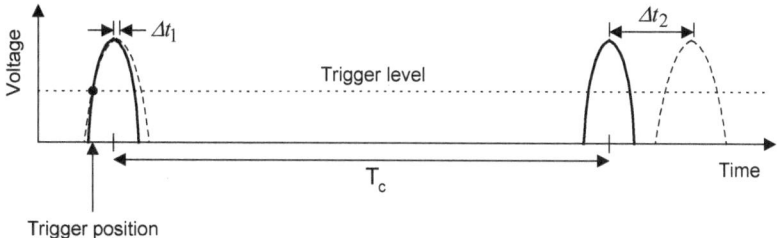

Fig. 6.5 Illustration of the monitoring problem in obtaining matching of the initial coupling signals included in the reference and the real signals

forms are completely overlapped. This part is actually the "initial coupling signal." This initial coupling signal always appears in all measurements due to the initial coupling of the transmitting signal into the receiving antenna regardless whether a target is present or not. The initial coupling signal can be used as a pilot signal in synchronization because it can be an indicator of the matching status of the two background signals included in the reference and the real signals. If both the initial coupling signals are not matched exactly in a measurement, implying that the background signals in the reference and the real signals are not identical, then the subtraction process may produce significant spurious signals after subtraction and hence may generate false alarms. To maintain matched initial coupling signals, the DAQ (or ADC) parameters such as the trigger slope and the trigger level should be maintained when measurements are carried out to obtain the reference and the real signals. Even though the DAQ parameters are maintained, mismatch between the initial coupling signals can occur because of the frequency variation in the reference clock oscillators. Small frequency modulation in the reference clock oscillator can cause the down-converted signal to be stretched or shortened in time scale, which is a frequency modulation effect on the down-converted signal. If this frequency modulation effect occurs in real signal measurement after recording the reference signal, there will be significant mismatch between the two background signals in the reference and the real signals. To compensate for this mismatch between the background signals, the control voltage of the VC-TCXO used in the UWB system should be adjusted to compensate for the frequency modulation effect. By monitoring the pilot signal, which is the initial coupling signal, we can detect the frequency modulation effect on the down-converted signal and compensate for it by adjusting the control voltage of the VC-TCXO. However, monitoring of the matching status in the short time interval of the initial coupling signal is very difficult and impractical because an insensible mismatch in the initial coupling signals could cause significant errors in the subtraction process. To overcome this monitoring problem, a revised monitoring method is used.

Figure 6.5 shows a simplified schematic representation of the problem in the monitoring of the matching status of the initial coupling signals. In Fig. 6.5, the pulse signals represented as a solid line is the initial coupling signal of the reference signal, and another pulse signal represented as a dotted line is the initial coupling signal of the real signal. One cycle of pulse signals with a standard clock period T_c is presented here. In ideal conditions, there is a time interval of a standard clock period T_c between two peak pulses for each signal. In other words, for the ideal matched condition of initial coupling signals, there is no frequency modulation in the reference clock so that the standard clock period T_c is maintained and the real signal coincides perfectly with the reference signal. However, if frequency modulation occurs, the situation depicted in Fig. 6.5 can appear, and the initial coupling pulses are not matched. As shown in Fig. 6.5, the deviation between the reference and the real signals is as small as Δt_1 in trigger position, which is the same situation as in Window-1 in Fig. 6.4. On the other hand, after one cycle of the clock, the amount of deviation Δt_2 is much larger than Δt_1, which means that monitoring of the matching status in a single clock period after the trigger position can easily identify small mismatch that may not be detected at the trigger position. Using this revised monitoring method, we can detect minor deviation in the background signals to be subtracted, compensate for the frequency modulation of the clock oscillator by adjusting its control voltage, and eventually obtain accurate reflected signals from the target only by background subtraction.

Now we consider Window-2 and Window-3 in Fig. 6.4. Window-2 represents the resulting data of background subtraction. This is a general A-scope display format typically used in radar systems for target signal detection. The data in Window-3 are simply the absolute value of the data in Window-2. Comparing the data in Window-2 and Window-3, we can see that the data in Window-3 alleviate some visual ambiguity in identifying the target reflected signal from the background clutter signals because the data of Window-3 have a single polarity rather than the dual polarity of the Window-2 data. The data format in Window-3 enables us not only to enhance the signal detectability but also to create the B-scan data format, which will be shown below.

In practical applications of the UWB systems involving assessment of stratified structures, the B-scan data format shown in Fig. 6.6 is commonly used to detect the target signatures reflected from the layer interfaces and to visualize the internal structure of the illuminated target structure. The horizontal axis shows the number of samples representing the depth information from the end of the antenna aperture and the vertical axis represents the number of scans, where a scan is defined as a measurement with selected pulse duration and at a certain location on the target structure. If the system is moving over a target structure for measurement, then the scan number corresponds to the displacement of the moving system with respect to a certain reference position on the target structure. The intensity level represents the amplitude of the detection data, defined as the amplitude level obtained from Window-3 of the detection program in Fig. 6.4. In Fig. 6.6, a darker intensity level represents higher amplitude of the returned signal, which is more likely a target reflected signal. Using the B-scan format, multiple scan data can be gathered into a single format and help us visualize the overall scan information. With UWB systems having multiple frequency bands such as the UWB system presented in this chapter, the B-scan format is especially useful for comparing all the detection data obtained from multiple frequency bands.

Fig. 6.6 Display of the
B-scan data format. The
horizontal axis represents
the number of samples. The
vertical axis represents the
number of scans. The time
interval between two adjacent
samples is the sampling interval, which is the inverse of
the sampling rate of the ADC

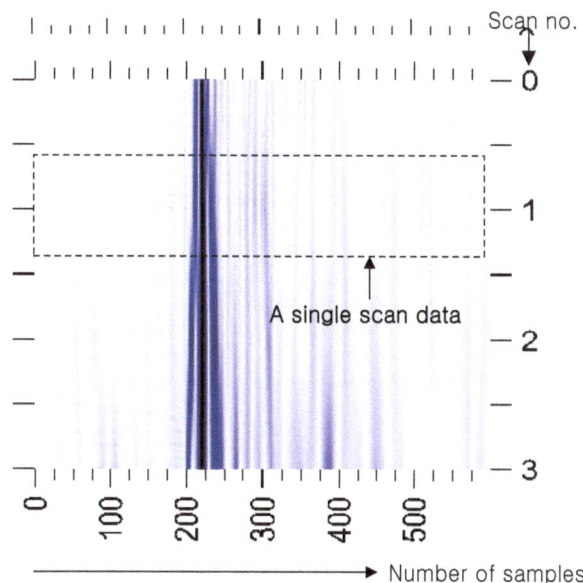

Fig.6.7 Photograph of the
UWB system prototype

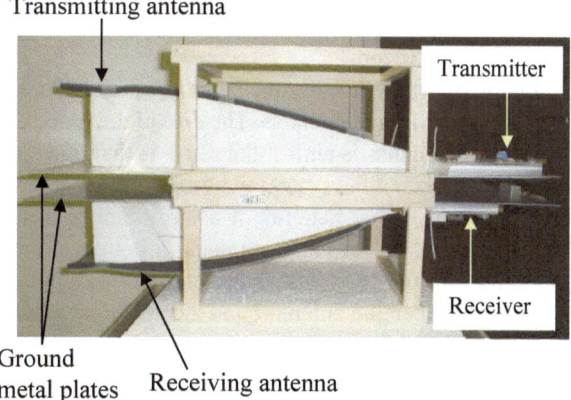

6.4   UWB System Integration

The designed UWB system consists of the UWB transmitter-antenna and receiver-
antenna integrated in a single structure, and the signal processing unit. Figure 6.7
shows a photograph of the UWB system using a wooden frame as a laboratory prototype for simple illustration of the complete system. For field operation, the UWB
system needs to be manufactured in a more rigid form. The signal processing unit
as described in Sect. 6.2 is not shown here. It should be noted that the size of this
UWB system can be reduced by reducing the size of the antenna and fabricating
the antenna and transmitter or receiver on a single substrate using all microstrip cir-

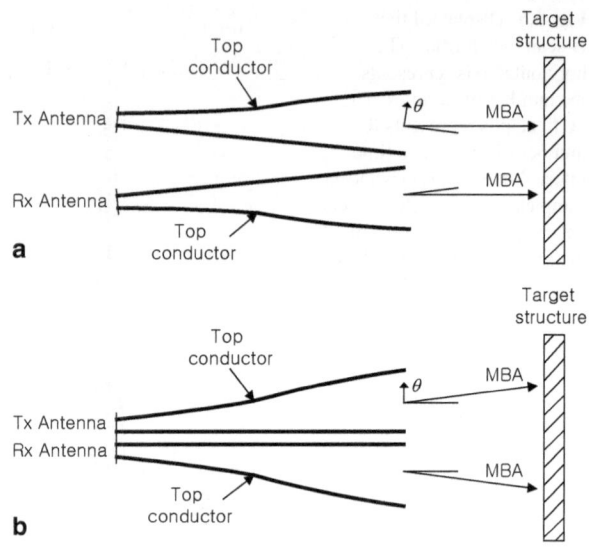

Fig. 6.8 Illustration of two possible antenna arrangements to achieve matching of the main beam axes (MBA) **a** and reduce cross coupling **b** between two antennas

cuits. Our calculations show that the transmitter-antenna or receiver-antenna can be easily fitted within a single substrate having an area of 5 in. (width) × 8 in. (length). That means the entire UWB system can have a size of only 5 in. (width) × 8 in. (length) × 6 in. (depth).

Figure 6.7 shows clearly the benefits of using the microstrip quasi horn antennas instead of TEM horn antennas. The size of the space occupied by the two microstrip quasi horn antennas is almost the same as that of a single TEM horn antenna, making the total size of the system to be significantly reduced from that using two TEM horn antennas. Other benefits of using the microstrip quasi horn antennas, as can be seen in Fig. 6.7, result from using a bottom ground metal plate, which can be the same for both the transmitting and receiving antennas, are the inherently high isolation between the antennas and a convenient place for mounting the transmitter and receiver with the antennas. Note that, in a complete microstrip environment, the transmitter and receiver designed on microstrip can be seamlessly integrated with the antennas on the same substrate above the common ground plane.

An important factor in the UWB system integration is the arrangement of the transmitting and receiving antennas. Figure 6.8 shows the sketch for two possible arrangements of the antennas. In Fig. 6.8a, the two antennas are arranged to tilt toward each other at 5-degree angles (in θ-axis) to match the main beam axis of these antennas according to the antenna's measured results shown in Table 5.2 of Chap. 5. This arrangement may result in maximum return signal power from the target, but the cross-coupling between the antennas is increased. The other arrangement in Fig. 6.8b shows the same structure as that used in the UWB system shown in Fig. 6.7, in which the two antennas are arranged to be parallel to each other instead of tilting toward each other at 5-degree angles. In this arrangement, the main beam axes of both antennas are deviated a little from each other, which results in less return signal power from the target, but decreased cross-coupling between the antennas may be obtained. Experimental results of the UWB system showed that the cross-coupled signals often produce spurious signal detection and increase in the false alarm rate when using the background subtraction technique in the signal processing. To reduce the spurious signal detection caused by the cross-coupling signal, the arrangement shown in Fig. 6.8b is used for the final UWB system.

6.5 Test and Evaluation of the UWB System

The designed UWB system was tested as a subsurface penetrating radar to illustrate its specific possible use for subsurface sensing as well as to demonstrate its workability and performance as a UWB system in general. Other kinds of tests can also be conducted for the UWB system for different applications such as detection and classification of buried mine and UXO, through-wall imaging and surveillance, locating and tracking people and goods, etc. In the following, we will show four different tests: one for a metal plate as a way to assess the working potential of the UWB system, one for stratified structures and another for a pavement sample to assess the performance for multilayer targets, and another for buried UXO (unexploded ordnance) to assess the UWB system's potential for detection and classification of buried UXO's and mine.

6.5.1 Test for a Metal Plate

A metal plate is the perfect reflector for electromagnetic pulses. Thus, a test using a metal plate is a rudimentary one typically executed first for a UWB system to evaluate its working potential through the pulse transmission to and reception from the metal plate. For this test, the metal plate is placed on the ground in an indoor laboratory, and the UWB system is placed 18 inches above the center of the metal plate. Figure 6.9 shows the pulses received from illuminating the metal plate with four different pulses having different pulse durations. The results show that all that detected waveforms do not have severe distortion in the main-lobe or serious ringing, except that corresponding to the 1170-ps transmitting pulse. The detected pulse from the 1170-ps transmitting pulse shows large ringing which is caused by multiple reflections from the indoor surrounding clutters. The occurrence of the large ringing implies that the designed antenna has such a large beam width that 1170-ps pulse cannot be focused on a spot on the object. It is clear from this result that the designed UWB system may not work well for 1170-ps pulse, or equivalently at very low frequencies, and hence the antenna needs to be optimized further for 1170-ps pulse operation.

Figure 6.10 shows the B-scan format of the detection data for the metal plate, in which each scan represents the data for a specific transmitting pulse duration.

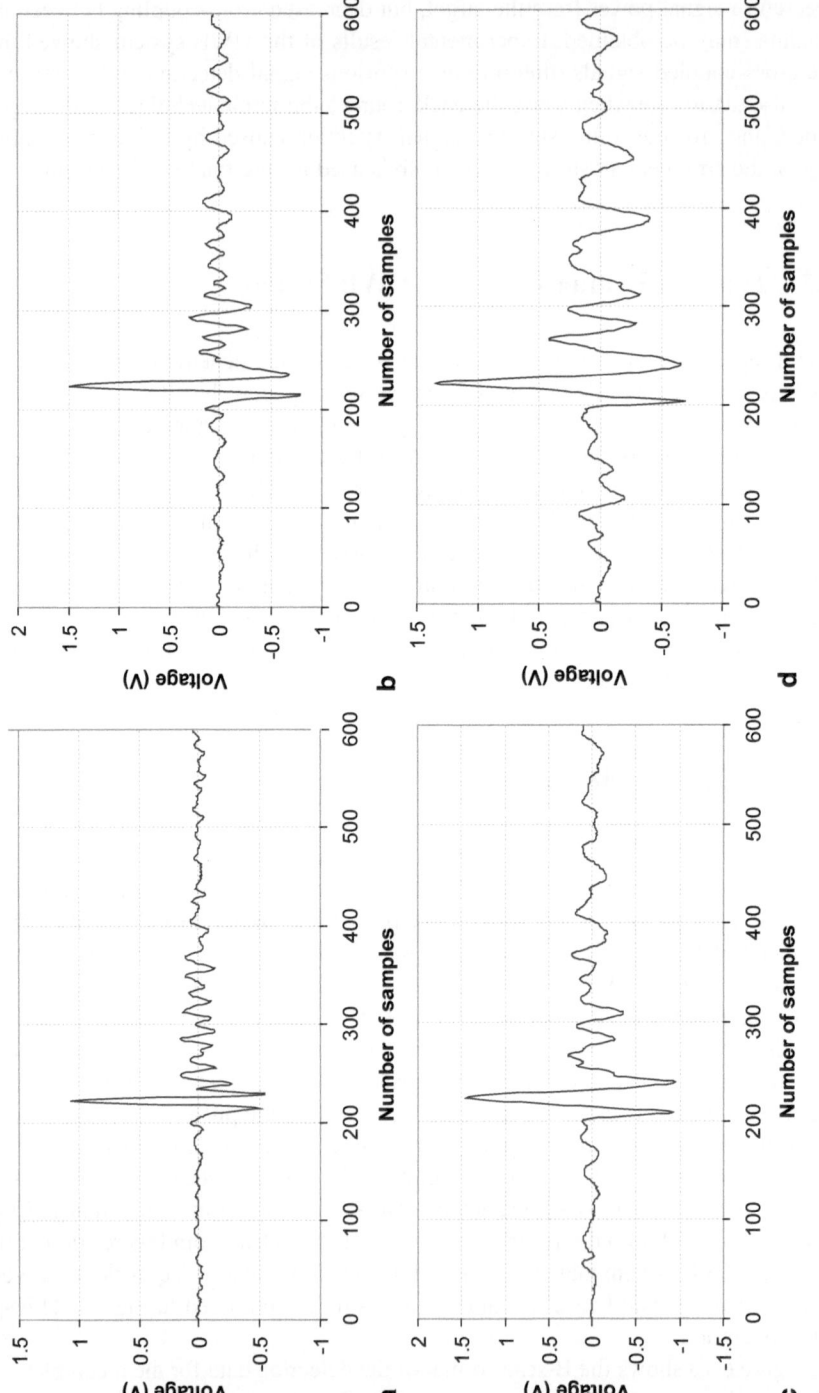

Fig.6.9 Received signals from a metal plate corresponding to transmitting signals of **a** 450 ps, **b** 600 ps, **c** 880 ps, and **d** 1170 ps

Fig. 6.10 B-scan format
results for received signals
reflected from a metal plate

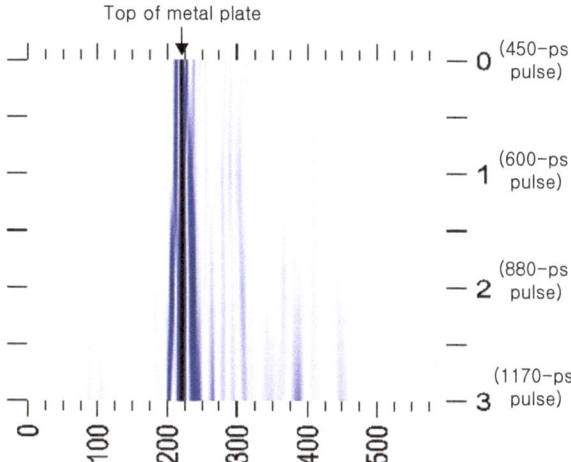

6.5.2 Tests for Stratified Structures

Figure 6.11 shows three sample structures used for assessing the UWB system performance in characterizing stratified structures. Sample A in Fig. 6.11a contains a
thin 1.375-in wood layer and is intentionally used for evaluating the range resolution
of the system. Sample B in Fig. 6.11b uses a much thicker wood layer to produce
the effect of a thick material on the system's detection performance. Sample C in
Fig. 6.11c has another wood layer embedded within the structure to demonstrate the
detection capability for a more complex stratified structure. In all measurements,
the UWB system was pointed directly onto the samples through air. Styrofoam,
having a relative dielectric constant of around 1, was used as the support. Wood is
known to have a very wide range of relative dielectric constant, from 1.2 to 5, and
loss tangent value ranged from 0.004 to 0.4167.

The assessment of the test structures in Fig. 6.11 is done in two steps. In the first
step, the reflected signals from the test structures are measured. In the second step,
the relative dielectric constant and thickness of each layer in the test structures are
determined based on the results obtained in Step 1.

6.5.2.1 Measurement of Reflected Signals

Figure 6.12 shows simultaneously the measured detection results for sample A for
four different pulses in the B-scan format. As can be seen, the three interfaces between the layers are clearly detected: air-wood interface (top of the wood), wood-
Styrofoam interface (bottom of the wood), and Styrofoam-ground interface (top of
the ground). These results show that a 450-ps pulse can definitely support about 1
inch of range resolution, which is close to what we expected. For the 600- and 880-
ps pulse, the system also detects the bottom of the wood, signifying that they can

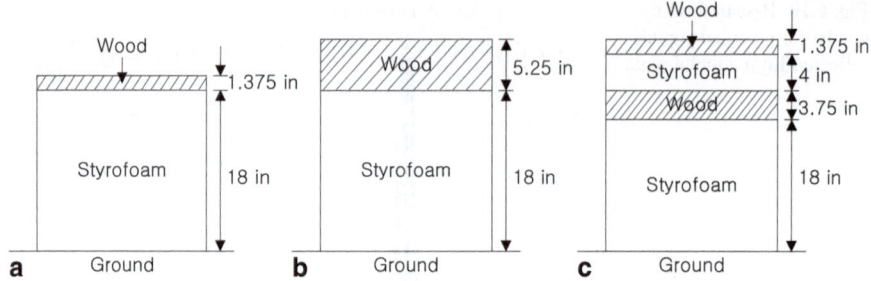

Fig. 6.11 Stratified structures: **a** sample A, **b** sample B, and **c** sample C

Fig. 6.12 Measured detection result for sample A

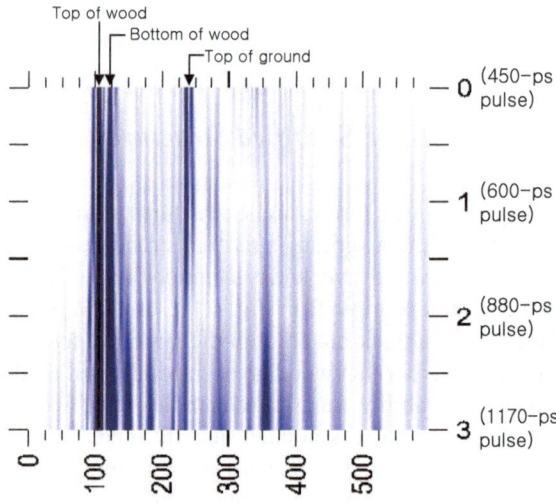

also support about 1 inch of range resolution. This range resolution for 600- and 880-ps pulses is finer than that estimated based on non-overlapped pulses discussed in Sect. 2.5 of Chap. 2. The reason of this finer-than-expected range resolution for large pulse durations will be explained later in detail.

Figure 6.13 shows the detection data for sample B. The system clearly detects the 2nd and 3rd interfaces, which are at the bottom of the wood layer and on top of the ground, respectively. Clutter signal between the signals reflected from the 2nd and 3rd interface, however, is also observed. This shows that target structures having layers with greater than 5-in thickness may produce some internal clutter signals. Nevertheless, in the case of 450- and 600-ps pulse durations, the clutter signal is relatively much weaker than the actual signals reflected from the interfaces.

Figure 6.14 shows the detection results for sample C. As in the previous cases, it shows clear detection of all interfaces between layers and the clutter signal between the 4th and 5th interface reflected signals.

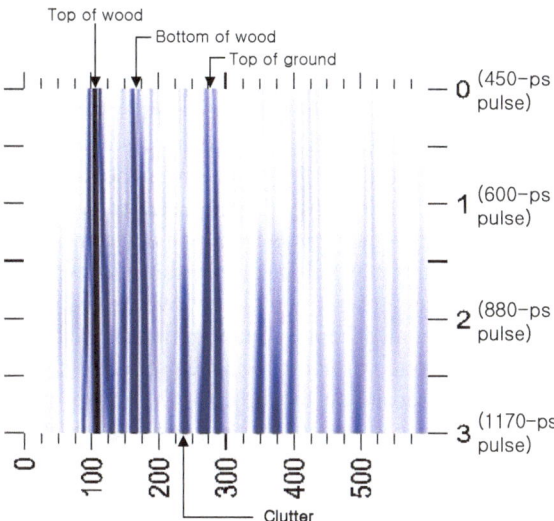

Fig. 6.13 Detection result for sample B

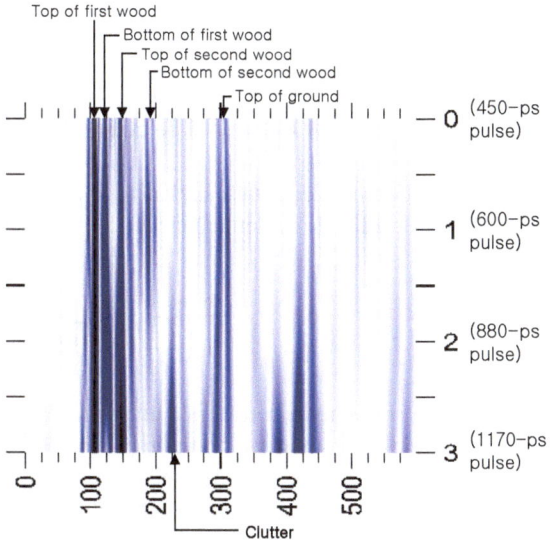

Fig. 6.14 Detection result for sample C

From the experimental results for these different types of sample structures, it is found that the 450- and 600-ps pulse durations support finer range resolution and reduced clutter signals than the 880- and 1170-ps pulse durations, and hence are more useful for the designed UWB system. Between the two shorter pulse durations, the 600-ps may provide a compromise to achieve a moderate penetration depth with relatively good range resolution.

In the detection results of Figs. 6.12 and 6.14, it is observed that the bottom of the first wood layer of about 1-inch thickness can be detected by using not only a

450-ps pulse, but also 600- and 880-ps pulse. It is recalled that the minimum pulse duration required to achieve 1 inch of range resolution is estimated as 400 ps in Sect. 2.5 of Chap. 2 using an asphalt layer as an example. The experimental detection results, however, show that even pulses of larger durations may also produce about 1-in range resolution. The main reason for this discrepancy between the estimated and measured range resolution can be found through the detection waveform shown in Fig. 2.9 of Chap. 2. The detected signal in Fig. 2.9 represents two non-overlapped reflected signals from two adjacent interfaces. We can see that the two reflected signals may still be discernable even when they come closer to each other, as long as they are not completely overlapped. This means that the actual range resolution obtained from a certain pulse may be better than the estimated one based on the assumption of non-overlapped reflected signals. This then indicates that the actual range resolution for the 600- and 880-ps pulses is better than the expected one and may even detect 1-in thickness. This better-than-estimated resolution fact is also confirmed in other measurement results.

6.5.2.2   Measurement of Relative Dielectric Constant and Thickness of Layers

From the detection results of the signals reflected from the test structures A, B and C presented in the foregoing section, the relative dielectric constant and thickness of each layer in these structures can be determined using the procedure introduced in [3].

For a stratified medium such as that as shown in Fig. 2.8, the relationship between the reflected and incident electric field intensity normal to the interface can be derived as

$$\frac{E_{rn}}{E_i} = \Gamma_{n,n-1} \left(\prod_{m=1}^{n-1} T_{m,m-1} T_{m-1,m} \; e^{-2\alpha_m d_m} \right) \tag{6.1}$$

where E_{rn} is the reflected electric field intensity from the nth layer interface, E_i is the incident electric field intensity on the top of the structure, $\Gamma_{n,\,n-1}$ is the reflection coefficient at the interface between the $(n-1)^{th}$ and n^{th} layer, $T_{m,\,m-1}$ is the transmission coefficient from the $(m-1)^{th}$ to m^{th} layer, α_m and d_m are the attenuation constant and the thickness of the mth layer, repectively. The reflection coefficient $\Gamma_{n,\,n-1}$ for normal incidence can be expressed in terms of the intrinsic impedances as

$$\Gamma_{n,n-1} = \frac{\eta_n - \eta_{n-1}}{\eta_n + \eta_{n-1}} \tag{6.2}$$

where η_n is the intrinsic impedance of the nth layer, which it can be approximated as a real number for low-loss and non-magnetic materials as

$$\eta_n \approx \frac{377\Omega}{\varepsilon_{rn}} \tag{6.3}$$

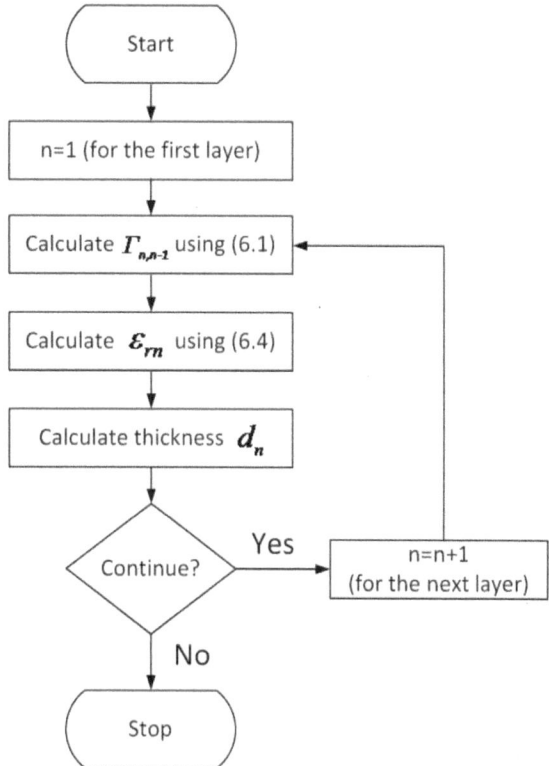

Fig. 6.15 Procedure for calculating the relative dielectric constant and thickness of layers in a multilayer target structure

with ε_{rn} being the relative dielectric constant of the nth layer. By substituting (6.3) into (6.2), we can derive an equation for ε_{rn} as

$$\varepsilon_{rn} = \varepsilon_{r,n-1} \left(\frac{1-\Gamma_{n,n-1}}{1+\Gamma_{n,n-1}} \right)^2 \qquad (6.4)$$

Figure 6.15 shows a flow-chart illustrating the procedure of calculating the relative dielectric constant and thickness of each layer of the multilayer structure based on Eqs. (6.1–6.4). The procedure is elaborated as follows considering an n-layer structure similar to that shown in Fig. 2.8 of Chap. 2.

The reflection coefficient at the interface between layers 0 and 1, Γ_{10}, can be determined from (6.1) as

$$\Gamma_{10} = \frac{E_{r1}}{E_i} \qquad (6.5)$$

It is recognized that, for a perfect metal-plate target, $\Gamma_{10} = E_{r1}/E_i = -1$, and hence $E_i = -E_{r1} = -E_{metal}$, where E_{metal} represents the signal reflected from the metal plate. Consequently, E_i can be determined from the amplitude of E_{metal} and Γ_{10} can be calculated as $-E_{r1}/E_{metal}$, which is the ratio of the voltage amplitudes of the reflected signals from the actual target structure and a metal plate. The relative dielectric constant of layer 1, ε_{r1}, can then be calculated from (6.4), with $\varepsilon_{r,0}$ being the relative dielectric constant of air, from which we can determine the thickness of layer 1, d_1, from $d_1 = v_p t_0$, where v_p is the phase velocity of the wave propagating in layer 1 and t_0 is the time interval between the two reflected signals from the upper and lower interface boundaries, which is the theoretical propagation time in air. Let us define a time interval t_d as the measured time interval in the time scale of the down-converted signal. There is a linear relationship between the two time interval values, t_0 and t_d, such that $t_0 = t_d \cdot TSF$, where TSF is the time scale-factor. The TSF can be determined by measuring two reflected signals from a metal plate placed at two different locations. To that end, two measurements for a metal plate placed in two distances with difference $d = 1.5$ inch were performed, from which t_d can be determined as $t_d = 9.04$ ms. The theoretical propagation time is calculated as $t_0 = d/c = 127$ ps, where c is the phase velocity in free space. Therefore, the measured TSF can be obtained as $t_0/t_d = 14.0487 \times 10^{-9}$.

In general, the thickness of the nth layer, d_n, can be derived as

$$d_n = v_{pn} t_{0n} = \frac{c}{\sqrt{\varepsilon_{rn}}} \cdot t_{dn} \cdot TSF \qquad (6.6)$$

where v_{pn}, t_{0n}, and t_{dn} correspond to layer n. Using (6.1) and the fact that $E_i = -E_{metal}$, the reflection coefficient at the interface between layers 1 and 2, Γ_{21}, can be expressed as

$$\Gamma_{21} = \frac{-\dfrac{E_{r2}}{E_{metal}}}{T_{10}\, T_{01}\, e^{-2\alpha_1 d_1}} \qquad (6.7)$$

The transmission coefficients from layers 0 to 1, T_{10}, and layers 1 to 0, T_{01}, can be easily determined from Γ_{10} using the following relationship:

$$\begin{aligned} T_{n,n-1} &= 1 + \Gamma_{n,n-1} \\ T_{n-1,n} &= 1 - \Gamma_{n,n-1} \end{aligned} \qquad (6.8)$$

In (6.7), the calculation of $-E_{r2}/E_{metal}$ is similar to the case of $-E_{r1}/E_{metal}$ in the Γ_{10} calculation, d_1 is the calculated value in the previous step, and α_1 is the attenuation constant of the first layer in Np/in.

Notice that this method requires the information for the attenuation constant of each layer in order to calculate the reflection coefficients. In fact, the prerequisite of this method is that we must know in advance the material type of each layer of the target structure and its attenuation constant. As indicated in the procedure shown in Fig. 6.15, after calculating Γ_{21}, we can obtain the relative dielectric constant, ε_{r2}, and thickness, d_2, of the second layer using (6.4) and (6.6), respectively.

Table 6.1 Measurement relative dielectric constant (ε_r) and thickness of samples' layers

		Assumed ε_r	Measured ε_r	Actual Thickness (inch)	Measured Thickness (inch)	Assumed tan δ	Assumed α (Np/in.)
Sample A	Layer 1: (Wood)	1.2–5	4.0	1.375	1.389	0.1	0.106
	Layer 2: (Styrofoam)	1.0	1.16	18.0	17.53	0.0	0.0
Sample B	Layer 1: (Wood)	1.2–5	4.0	5.25	4.7	0.1	0.106
	Layer 2: (Styrofoam)	1.0	1.16	18.0	17.2	0.0	0.0
Sample C	Layer 1: (Wood)	1.2–5	4.0	1.375	1.43	0.1	0.106
	Layer 2: (Styrofoam)	1.0	1.22	4.0	3.52	0.0	0.0
	Layer 3: (Wood)	1.2–5	4.47	3.75	3.24	0.1	0.106
	Layer 4: (Styrofoam)	1.0	0.9	18.0	19.2	0.0	0.0

Table 6.1 summarizes the measurement results for the relative dielectric constants and thicknesses of the layers of samples A, B and C shown in Fig. 6.11 using the detected signals following the foregoing procedure. The results show good performance with small error and good signal detection for the developed UWB system. It is noted that, in Table 6.1, the assumed loss tangent of wood is chosen from known typical values in the range of 0.004–0.4167.

6.5.3 Test for Pavement

Figure 6.16 shows the profile and photograph of a pavement sample constructed for testing the UWB system as a subsurface penetrating radar. It is noted that a typical real pavement would consist of asphalt, base, and sub-base layers. However, in the pavement sample shown in Fig. 6.16, a structure comprising wood, air (void), and wood on ground is used in place of a sub-base. This is a reasonable replacement and, in fact, needed for assessing the performance of the UWB system in real applications, because, even in real pavement structures, air voids are occasionally encountered, and the detection of air voids is important in pavement assessment using subsurface penetrating radar. Typical values of the relative dielectric constants and other physical values for the asphalt and base are given in Tables 2.1 and 2.2 of Chap. 2.

Figure 6.17 shows the detection results for the pavement sample of Fig. 6.16. The detected signals are the reflected signals from all of the major interfaces in the pavement sample, which are the top of the asphalt, the asphalt-base interface (or the top of the base), the bottom of the base and the wood-air interface, and the air-wood interface. The third detected signal indicates that the two interfaces, which are the bottom of the base and the wood-air interface, can not be clearly distinguished. It is also noticed that the top of the base is only detected by the 450- and 600-ps

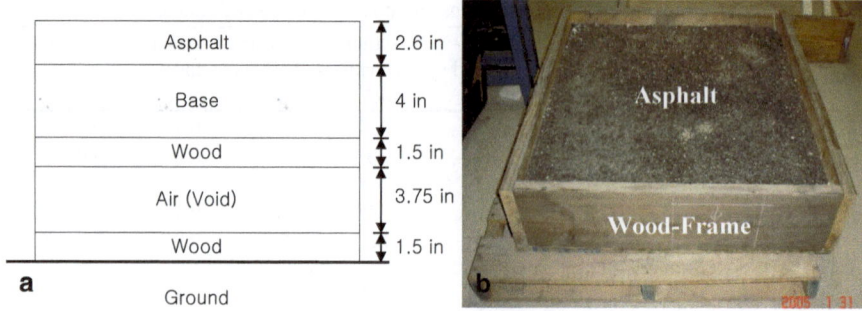

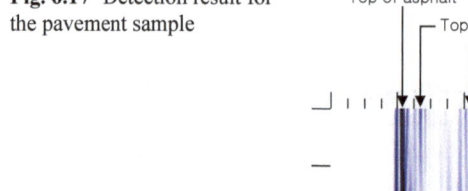

Fig. 6.16 Profile **a** and photograph **b** of a pavement sample

Fig. 6.17 Detection result for the pavement sample

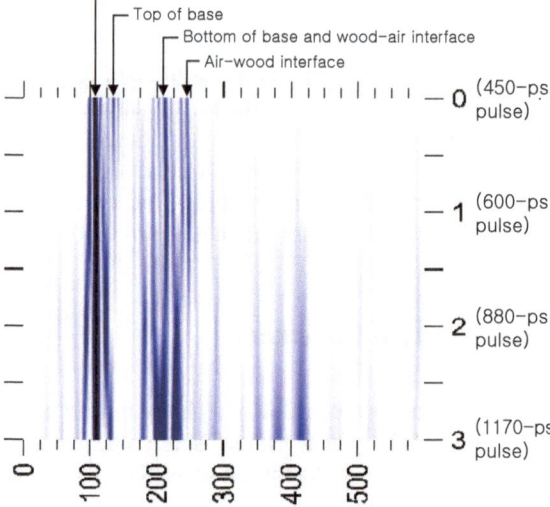

monocycle pulses. From these results, it can be concluded that the 450- and 600-ps pulses may be the most useful pulse signals for pavement assessment using the designed UWB system, which is similar to what is found in the previous results for the sample structures with wood layers shown in Fig. 6.11.

Table 6.2 shows the measurement results for the relative dielectric constant and thickness of the asphalt and base of the pavement sample shown in Fig. 6.16 using the detected signals. The results show the accuracy of the measurement as well as good detection capability of the developed UWB system for pavement assessment, similar to the assessment of the stratified media described in Sect. 6.4.2. Note that in Table 6.2, the assumed loss tangent value of the asphalt is chosen from known typical values in the range of 0.004 to 0.01.

Table 6.2 Measured relative dielectric constant and thickness of asphalt and base layer of the pavement sample

	Assumed ε_r	Measured ε_r	Actual Thickness (inch)	Measured Thickness (inch)	Assumed tan δ	Assumed α (Np/in.)
Layer 1: (Asphalt)	5–7	3.72	2.6	2.376	0.004	0.006
Layer 2: (Base)	8–12	7.0	4.0	4.234		

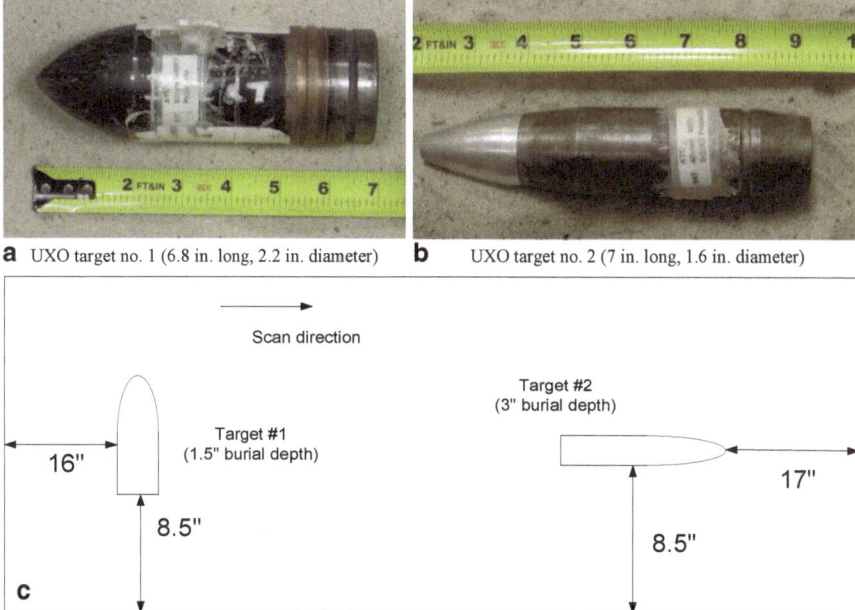

a UXO target no. 1 (6.8 in. long, 2.2 in. diameter)   **b**   UXO target no. 2 (7 in. long, 1.6 in. diameter)

Fig. 6.18 Two UXO targets (**a**, **b**), and burial profile in a sand box (**c**)

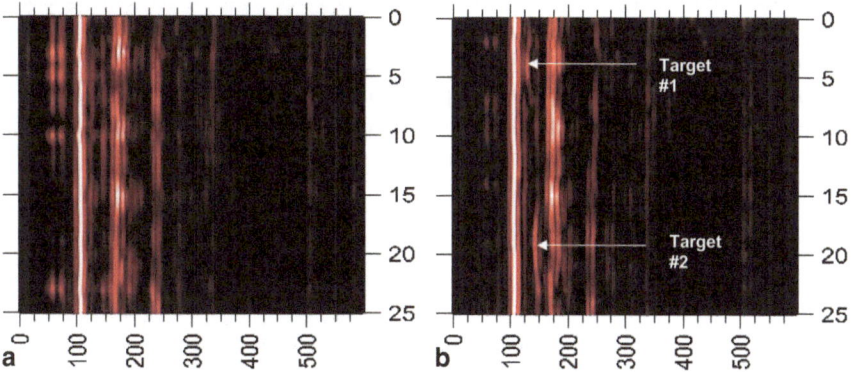

Fig. 6.19 Scanning results of a sand box without **a** and with **b** buried UXO targets

6.5.4   UXO Test

As an example demonstrating the designed UWB system's potential for detection and classification of buried UXO's and mine, we conducted some simple experiments for detection of various UXO's buried under different soils and depths. Figure 6.18 shows two UXO targets buried under sand.

Figure 6.19 shows the scanning results of the bare sand without the buried UXO's (for reference) and buried UXO targets being illuminated by a 400-ps monocycle pulse from the UWB system. As can be seen, the two targets can be detected and their shapes and burial depths can be identified. It is noted that these are raw data without any signal processing. It is expected that with good signal processing, much better information of the targets can be extracted.

6.6   Summary

The development of a UWB system has been presented. The system integrates the transmitter, receiver and antennas in a single package using microwave integrated circuits. It possesses unique capability of tunable transmitting pulses, effectively simulating a multi-band UWB system or combination of multiple UWB systems. The system is capable of transmitting monocycle pulses having duration varied from 450 to 1170 ps and peak-power ranged from 200–400 mW, and detecting target signals with more than 50-dB dynamic range and 6.5–9.5 dB conversion gain over a 5.5-GHz RF bandwidth. It achieves a resolution as small as 1 inch. Various tests were carried out to demonstrate not only the performance of the developed system, but also its validity as subsurface penetrating radar. Test results for stratified and pavement structures show that the 450- and 600-ps pulses render low clutter signal levels and better range resolutions among the available pulses for the system. The multi-pulse/multi-band feature of the system provides flexibility in the operation and use of the system for various applications, achieving not only fine range resolution but also deep penetration (or long range) for sensing. Another potential advantage of multiple pulse durations is the possibility of enhanced target detection and classification due to more information obtained from targets. The developed system should also be useful for other non-subsurface-sensing applications such as through-wall imaging and surveillance for locating and tracking individuals within buildings, building surveillance and monitoring, etc., and detection of concealed weapons.

References

1. Daniels, D.J.: Short pulse radar for stratified lossy dielectric layer measurement. IEE Proc. 127(5), 384–388 (October 1986)
2. LabVIEW User Manual. National Instruments, Austin (1998)
3. Park, J.S., Nguyen C.: An ultra-wideband microwave radar sensor for nondestructive evaluation of pavement subsurface. IEEE Sens. J. 5, 942–949. (October 2005)

Chapter 7
Summary and Conclusion

In the last six chapters, this book has covered the theory, analysis, and design of UWB systems. Specifically, the book addresses the analysis, design, integration, and test of a UWB system capable of operation with multiple varying-duration pulses and its constituent components and subsystems including transmitter, receiver and antenna. This UWB system integrates the transmitter, receiver, antennas and signal processing in a single unit and can be used for various sensing applications.

In the first chapter, the book introduces the UWB systems with essential characteristics and vast applications.

The second chapter addresses the theory and analysis of the UWB systems, particularly the system operation, UWB pulse signals, power budget, and range resolution. An example showing the calculations for minimum required power and pulse duration corresponding to a particular range resolution for the transmitting signal is also presented for a typical pavement sample.

The third chapter presents the design of a printed-circuit UWB transmitter using microwave integrated circuits (MICs) that can produce multiple monocycle pulses of different durations. The UWB transmitter can generate four different monocycle pulse durations of 450, 600, 880, and 1170 ps, which are the UWB signals covering the frequency range from 0.15 to 3.7 GHz. The monocycle pulse duration is reconfigurable by simply changing the position of the employed p-i-n diodes. The UWB transmitter circuit can generate relatively high pulse peak powers of 200–400 mW for 50-Ω load using a low DC bias voltage.

In the fourth chapter, the design of a printed-circuit UWB receiver implementing the synchronous sampling is described. The UWB receiver integrates a coupled-slotline-hybrid sampling mixer, local strobe pulse generator, and baseband video amplifier in a single printed-circuit board using MICs. The UWB receiver has a conversion loss of 4.5–7.5 dB (without amplifier) and conversion gain from 6.5–9.5 dB (with amplifier) over a 3-dB bandwidth of 5.5 GHz. It has a dynamic range greater than 50 dB and a sensitivity of -47 dBm. The UWB receiver can reproduce an (input) signal faithfully, matching that measured by a commercial digitizing oscilloscope.

C. Nguyen, J. Han, *Time-Domain Ultra-Wideband Radar, Sensor and Components*, SpringerBriefs in Electrical and Computer Engineering, DOI 10.1007/978-1-4614-9578-9_7, © Springer International Publishing Switzerland 2014

The fifth chapter covers the design of two UWB antennas: the UWB microstrip quasi-horn antenna and the UWB uniplanar antenna. Both antennas were tested in frequency- and time-domain and exhibit good performance under operations using multiple different pulses of various durations. The UWB microstrip quasi-horn antenna is particularly used for the UWB system.

The sixth chapter addresses the integration and tests of the UWB system that comprises the designed UWB transmitter, receiver, microstrip quasi-horn antennas, and signal processing unit, all packaged in a single unit. The tests were done through various target structures including metal plate, stratified media, pavement sample, and buried UXO's. The measurement results demonstrate the workability and good performance of the UWB system for the specific application of subsurface sensing and for potential applications in various other non-subsurface sensing. The test results also show that 450- and 600-ps monocycle pulses are the most useful ones among available pulses as they render low clutter signal levels and better range resolutions.

Although the UWB transmitter, receiver and antennas developed in this book are used specifically for the UWB system, they can also be used for other UWB time-main applications as well as in other UWB systems. Particularly, their analyses and design techniques would enable engineers to design UWB components and systems for a variety of specifications and applications. Moreover, the developed UWB system can be used for many other UWB applications.

It is noted that the UWB system presented in this book is not intended for field operations. Rather, it is a simple laboratory prototype used to demonstrate the design, anlaysis, and applications of UWB systems. As such, the developed UWB system is not optimum, both physically and electrically. Nevertheless, it is relatively straightforward to produce a working system suitable for field operations including better manufacturing for a more physically rigid system as well as substantial size reduction to make it portable.

Lastly, we wish to emphasize that, although this book is relatively concise, it contains sufficient practical and valuable information that should enable the readers to successfully design UWB components, transmitters, receivers, antennas, and systems relatively easy for their own use in many applications.

Index

C. Nguyen, J. Han, *Time-Domain Ultra-Wideband Radar, Sensor and Components,*
SpringerBriefs in Electrical and Computer Engineering,
DOI 10.1007/978-1-4614-9578-9, © Springer International Publishing Switzerland 2014